KT-145-420

Digital Electronic Technology

A textbook covering the digital electronic
content of the Level IV syllabus of the
Technician Education Council

D C Green
M Tech, CEng, MILRE
Senior Lecturer in Telecommunication Engineering
Willesden College of Technology

Pitman

PITMAN BOOKS LIMITED
128 Long Acre London WC2E 9AN

Associated Companies
Pitman Publishing New Zealand Ltd, Wellington
Pitman Publishing Pty Ltd, Melbourne

© D. C. Green 1982

First published in Great Britain 1982

All rights reserved. No part of this publication may be reproduced,
stored in a retrieval system, or transmitted, in any form or by any
means, electronic, mechanical, photocopying, recording and/or
otherwise without the prior written permission of the publishers. This
book may not be lent, resold, hired out or otherwise disposed of by
way of trade in any form of binding or cover other than that in which it
is published, without the prior consent of the publishers. The book is
sold subject to the Standard Conditions of Sale of Net Books and may
not be resold in the UK below the net price.

Filmset and printed in Northern Ireland at The Universities Press (Belfast) Ltd.

ISBN 0 273 01722 5

Preface

This book provides a comprehensive coverage of the more important circuits and techniques used in modern digital electronic circuitry. The various logic families are discussed and their relative merits considered but the description of the digital circuits, such as counter, is illustrated by reference to the two most popular families, i.e. ttl and cmos, only.

The amount of desirable knowledge for an electronic/telecommunication technician or engineer is very large and greater than could possibly be tackled in a standard 60 hour unit. This means, of course, that some selection of the material to be covered by a particular course is necessary. The Technician Education Council (TEC) has produced a large unit for Level IV Electronics that covers most aspects of both analogue and digital electronics from which colleges are invited to select material to make up a 60 hour unit.

This book has been written to cover *all* the material specified in the digital sections of this large unit. The analogue sections are covered in the companion volume *Electronics* IV. Also included in this book is the section of the unit that refers to photo-electric devices which are used in both analogue and digital circuitry.

This book has been written on the assumption that the reader will already possess a knowledge of Electronics and of Digital Electronics up to the standard reached by the level III units of the TEC. The book should be useful to all students of digital electronics at a level that is approximately that of the TEC level IV unit given at the back of this book.

Acknowledgement is made to the Technician Education Council for its permission to use the content of its unit; the Council reserve the right to amend the content of its unit at any time.

D.C.G.

The following abbreviations for other titles in this series are used in this book:

[EII]   Electronics II
[EIII]   Electronics III
[DT&S]   Digital Techniques and Systems
[EIV]   Electronics IV

Contents

iv

1 Combinational Logic

Introduction

The **electronic gate** is a circuit that is able to operate on a number of binary signals in order to perform a particular logical function. The types of gate available to the designer are the AND, OR, NAND, NOR, exclusive-OR, and the exclusive-NOR (or the coincidence gate). Except for the exclusive-NOR gate, they can be obtained in monolithic integrated circuit form and practical examples will be discussed in the next chapter. The British Standard symbols for these gates are given in Fig. 1.1. Note that the & sign for the AND and the NAND gates can be replaced by a number indicating the number of inputs that must go high to make the output go high or low. Many logical functions can also be implemented by means of either a medium-scale integrated circuit device known as a *multiplexer* or a *read only memory* (rom).

Fig. 1.1 Gate symbols

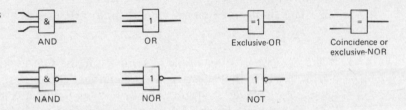

The operation of a logic circuit can be described by means of its *truth table*. This is a table which shows the output state of the circuit for all the possible combinations of the input variable states. The truth table of a

The Basic Gates

1 The **AND** gate is a logic element having two or more input terminals and only one output terminal. Its output is at logical 1 *only* when *all* of its inputs are at logical 1. If any one or more of the inputs are at logic 0, the output of the gate will also be at 0. The output F of a 4-input AND gate can be expressed using Boolean algebra as

$$F = A.B.C.D \tag{1.1}$$

The symbol for the AND logical function is the dot as shown but the dot is usually omitted, and the AND function is expressed as

$$F = A B C D$$

Each of the four input variables A, B, C and D is termed a *literal*.

The operation of a logic circuit can be described by means of its *truth table*. This is a table which shows the output state of the circuit for all the possible combinations of the input variable states. The truth table of a

Table 1.1 AND gate truth table

A	0	1	0	1	0	1	0	1	0	1	0	1	0	1	0	1
B	0	0	1	1	0	0	1	1	0	0	1	1	0	0	1	1
C	0	0	0	0	1	1	1	1	0	0	0	0	1	1	1	1
D	0	0	0	0	0	0	0	0	1	1	1	1	1	1	1	1
F	0	0	0	0	0	0	0	0	0	0	0	0	0	0	0	1

Table 1.2 OR gate truth table

A	0	1	0	1	0	1	0	1	0	1	0	1	0	1	0	1
B	0	0	1	1	0	0	1	1	0	0	1	1	0	0	1	1
C	0	0	0	0	1	1	1	1	0	0	0	0	1	1	1	1
D	0	0	0	0	0	0	0	0	1	1	1	1	1	1	1	1
F	0	1	1	1	1	1	1	1	1	1	1	1	1	1	1	1

4-input AND gate is given by Table 1.1. Note that the number of terms required is $2^4 = 16$.

2 The Boolean equation for a 4-input **OR** gate is given by

$$F = A + B + C + D \tag{1.2}$$

The + symbol is the Boolean symbol for the OR logical function. Its truth table is given in Table 1.2. It is clear that the output of an OR gate is at logic 1 whenever any one, or more, of its inputs are in the logical 1 state.

3 The **NAND** logical function is equivalent to the AND followed by an inversion, and hence the output of a NAND gate is at logic 0 *only* when *all* of its inputs are at 1. Otherwise its output is 1. This action is illustrated by the Boolean equation for a 4-input NAND gate (eqn. (1.3)) and by the truth table (Table 1.3).

$$F = \overline{A\,B\,C\,D} \tag{1.3}$$

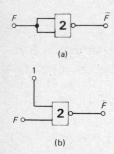

(a)

(b)

Fig. 1.2 The NAND gate as an inverter

A NAND gate can be used to provide the NOT, or inversion, function by connecting all of its input terminals together (Fig. 1.2*a*) or by connecting all but one of its inputs to logic 1 (Fig. 1.2*b*). The NAND gate is one of the most commonly employed small-scale integrated circuits and it is frequently used to produce both the AND and the OR functions. This is demonstrated later (p. 14).

4 The other commonly employed gate is the **NOR** gate. The truth table of a 4-input NOR gate is given by Table 1.4. This table should make it clear that the NOR gate gives the NOT OR logic operation. This means that its output F is at logical 1 only when *all* of its inputs are at logical 0. The NOR gate can also be used to produce the AND and the OR logic functions and can be used as an inverter (see Figs 1.3*a* and *b*). Either the two input

Table 1.3 NAND gate truth table

A	0	1	0	1	0	1	0	1	0	1	0	1	0	1	0	1
B	0	0	1	1	0	0	1	1	0	0	1	1	0	0	1	1
C	0	0	0	0	1	1	1	1	0	0	0	0	1	1	1	1
D	0	0	0	0	0	0	0	0	1	1	1	1	1	1	1	1
F	1	1	1	1	1	1	1	1	1	1	1	1	1	1	1	0

Table 1.4 NOR gate truth table

A	0	1	0	1	0	1	0	1	0	1	0	1	0	1	0	1
B	0	0	1	1	0	0	1	1	0	0	1	1	0	0	1	1
C	0	0	0	0	1	1	1	1	0	0	0	0	1	1	1	1
D	0	0	0	0	0	0	0	0	1	1	1	1	1	1	1	1
F	1	0	0	0	0	0	0	0	0	0	0	0	0	0	0	0

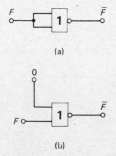

(a)

(b)

Fig. 1.3 The NOR gate as an inverter

terminals are connected together or one of the two inputs is connected to logic 0.

An alternative to the use of a NAND or a NOR gate as an inverter is to use a hex (six) inverter integrated circuit e.g. the ttl 7404.

5 The **exclusive-OR** gate has two input terminals and performs the logical function

$$F = A\,\bar{B} + \bar{A}\,B \tag{1.4}$$

The truth table describing this function is given by Table 1.5. The output of the gate is at logical 1 only when either one but *not* both of its inputs is at logical 1. If both inputs are at 0, *or* at 1, the output of the gate will be at logical 0.

The exclusive-OR gate can be purchased as an integrated circuit package or can be made up from a suitable combination of AND, NOT, and OR gates (Fig. 1.4a) or by using NAND gates only (Fig. 1.4b).

Commercially available ic packages only include 2-input exclusive-OR gates. If the functions

$$F = A \oplus B \oplus C \quad \text{or} \quad F = A \oplus B \oplus C \oplus D$$

are required, then two or three exclusive-OR gates can be connected in the manner shown by Fig. 1.4c. (The symbol $\oplus$ denotes the exclusive-OR function.)

Table 1.5 Exclusive-OR truth table

A	0	1	0	1
B	0	0	1	1
F	0	1	1	0

Fig 1.4 Exclusive-OR gate (*a*) using AND, OR, and NOT gates, (*b*) using NAND gates only, (*c*) for three or four literals

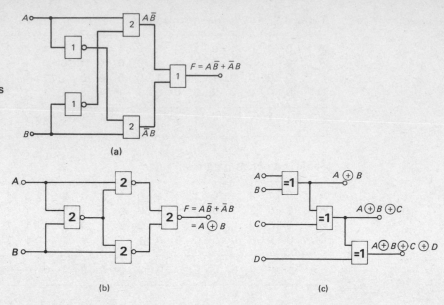

6 The last type of gate, the **exclusive-NOR,** is less often used than any of the other types. It is also known as the **coincidence gate** since its output is at logical 1 *only* when *both* of its inputs are at the *same* logical state, be it 0 or 1. The truth table of a coincidence gate is given by Table 1.6. From this table it is evident that the Boolean equation describing the coincidence gate is

Table 1.6 Exclusive-NOR truth table

A	0	1	0	1
B	0	0	1	1
F	1	0	0	1

$$F = \bar{A}\,\bar{B} + A\,B \tag{1.5}$$

It should also be observed that the output F of the gate is always the inverse of the output of the exclusive-OR gate. It is for this reason that the circuit is also known as the exclusive-NOR gate. The exclusive-NOR gate can be achieved in three different ways:

1) directly, using AND, NOT, and OR gates
2) by inverting the output of an exclusive-OR gate
3) using an *AOI gate* (p. 27) as shown by Fig. 1.5.

Fig. 1.5 Exclusive-NOR function generated by an AOI gate

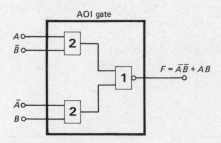

Boolean Equations   In the design of a combinational logic circuit, the truth table of the required operation can be written down and then used to derive the Boolean equation that expresses the output of the circuit in terms of the literals. The equation thus obtained can generally be simplified to reduce the number of gates required for its implementation. Very often the implementation is carried out using either NAND or NOR gates exclusively.

Boolean equations are often in either the *sum-of-products* form, e.g.

$$F = A\bar{B} + \bar{A}B$$

or the *product-of-sums* form, e.g.

$$F = (A + \bar{B})(\bar{A} + B)$$

In fact, any expression can be expressed in either of these forms, and one or the other may be the simpler to implement. A sum-of-products equation is often called a *minterm*, while a product-of-sums equation is sometimes known as a *maxterm*.

Canonical Form

The **canonical form** of a Boolean equation is one in which *each* term contains *all* of the literals *once* only. If two equations are written down in their canonical forms, they can be compared term by term and this fact will be utilized later when a tabular method of simplifying Boolean equations is described.

Example 1.1

Write the equation $\bar{A}B + \bar{C}B$ in its canonical form.

Solution

$$
\begin{aligned}
F &= \bar{A}B\,(C + \bar{C}) + \bar{C}B\,(A + \bar{A}) \\
&= \bar{A}BC + \bar{A}B\bar{C} + AB\bar{C} + \bar{A}B\bar{C} \\
&= \bar{A}BC + \bar{A}B\bar{C} + AB\bar{C} \quad (Ans)
\end{aligned}
$$

The Simplification of Boolean Equations

When the Boolean equation describing the logic operation of a circuit has been obtained, an attempt is always made to simplify, or **minimize,** the equation.

An equation is said to be minimized when it *a*) contains the lowest possible number of literals and *b*) the lowest possible number of terms. The minimized equation will usually require the fewest number of gates possible for its implementation, although this may not necessarily also give the least number of ic packages. Further, the minimal solution may well be subject to *race-hazards*.

There are three main methods by which a Boolean equation can be simplified; these are the use of Boolean algebra, the use of a Karnaugh map, and the use of tabulation techniques.

The **algebraic simplification** of logic functions is made easier by the use of the logic rules which follow:

1. $A + \bar{A} = 1$
2. $A + A = A$
3. $AA = A$
4. $A\bar{A} = 0$
5. $A + 0 = A$
6. $A + 1 = 1$
7. $A.1 = A$
8. $A.0 = 0$
9. $AB = BA$ ⎫
10. $A + B = B + A$ ⎬ Commutative law
11. $A(B + C) = AB + AC$ ⎫
12. $A + (BC) = (A + B)(A + C)$ ⎬ Distributive law
13. $A + B + C = (A + B) + C = A + (B + C)$ ⎫
14. $ABC = A(BC) = (AB)C$ ⎬ Associative law
15. $A(B + \bar{B}) = A$
16. $A + AB = A$
17. $A(A + B) = A$
18. $A + \bar{A}B = A + B$
19. $B(A + \bar{B}) = AB$
20. $(A + B)(B + C)(C + \bar{A}) = (A + B)(C + \bar{A})$
21. $\overline{A + B} = \bar{A}\bar{B}$
22. $\overline{AB} = \bar{A} + \bar{B}$

21 and **22** are known as **De Morgan's rules.** The accuracy of any of these rules can be confirmed with the aid of a truth table.

Example 1.2

Simplify $F = \bar{A}(B + \bar{C})(A + \bar{B} + C)\bar{A}\bar{B}\bar{C}$

Solution
Multiplying out gives

$$F = (AB + B\bar{B} + BC + A\bar{C} + \bar{B}\bar{C} + C\bar{C})\bar{A}\bar{A}\bar{B}\bar{C}$$

Rules 3 and 4 give

$$F = (AB + BC + A\bar{C} + \bar{B}\bar{C})\bar{A}\bar{B}\bar{C}$$

Multiplying out again and applying rules 3 and 4 gives

$$F = \bar{A}\bar{B}\bar{C} \quad (Ans)$$

Example 1.3

Simplify $F = (A + B)(\overline{AB} + C) + AB$

Solution
Rule 22 gives

$$F = (A + B)(\bar{A} + \bar{B} + C) + AB$$

Multiplying out and using rule 4 gives

$$F = A\bar{B} + AC + \bar{A}B + BC + AB$$
$$= A(B + \bar{B}) + AC + \bar{A}B + BC$$

$= A(1 + C) + \bar{A}B + BC$	from rule 1
$= A + \bar{A}B + BC$	from rule 6
$= A + B + BC$	from rule 18
$= A + B(1 + C)$	from rule 11
$= A + B$ (*Ans*)	from rule 6

With some practice it is possible to write several of the above steps down at once.

The algebraic method of simplifying Boolean equations possesses the disadvantages that *a*) it can be very time consuming and *b*) it requires considerable practice and experience before the most appropriate approach and/or rule can be selected quickly. Usually it is better to employ the mapping method described later.

Duality

Every sum-of-products equation has a dual product-of-sums equation and vice versa, and this fact provides another method which can be used in the simplification of Boolean equations.

The **dual** of an equation is simply obtained by merely replacing every AND symbol (.) by the OR symbol (+) and vice versa. For example,
the dual of $F = (\bar{A} + B + C)(A + D)$ is $F' = \bar{A}BC + AD$

Example 1.4

Use duality to simplify Example 1.3

Solution

$$F = (A + B)(\bar{A} + \bar{B} + C) + AB$$
$$F' = (AB + \bar{A}\bar{B}C)(A + B)$$
$$= AB$$

Therefore $F = A + B$ (*Ans*)

Clearly in this case the algebra involved is much simpler than previously.
The complement $\bar{F}$ of a Boolean equation can be obtained by replacing each literal by its complement in the corresponding dual equation.
Thus if $F = AB + C$, then $F' = (A + B)C$ and so

$$\bar{F} = (\bar{A} + \bar{B})\bar{C} = \bar{A}\bar{C} + \bar{B}\bar{C}$$

The Karnaugh Map

The Karnaugh map provides a convenient method of simplifying Boolean equations in which the function to be simplified is displayed diagrammatically on a map of squares. Each square maps one term of the function. The number of squares is equal to 2^n, where n is the number of literals in the function. Thus, if the equation to be simplified has three literals A, B and C, then $n = 3$ and $2^3 = 8$, and so 8 squares are needed. The rows and columns of the Karnaugh map can be labelled as shown for 4, 8 and 16 square maps. (The labels *in* the squares are not normally given.)

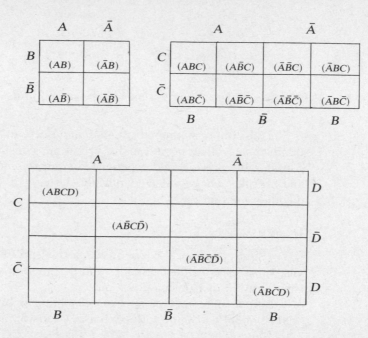

The number 1 written in a square indicates the presence, in the function being mapped, of the term represented by that square. The number 0 written in a square means that the particular term is not present in the mapped function.

To simplify an equation using the Karnaugh map adjacent squares containing 1 are looped together. This step eliminates any terms of the form $A\bar{A}$.

In this context, *adjacent* means *a*) side-by-side in the horizontal and in the vertical directions (but *not* diagonal), *b*) the right-hand and left-hand sides, and the top and bottom, of the map and *c*) the *four* corner squares. Squares can be looped together in two, fours or eights *only*.

Example 1.5

Use a Karnaugh map to simplify the equation

$$F = ACD + \bar{A}BCD + \bar{B}\bar{C}\bar{D} + \bar{A}\bar{B}C\bar{D} + \bar{A}\bar{B}CD + A\bar{B}C\bar{D}$$

Solution
The mapping of the equation is shown.

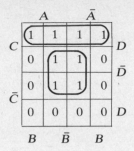

Looping the adjacent squares in two groups of four simplifies the equation to

$$F = CD + \bar{B}\bar{D} \quad (Ans)$$

When $\bar{F}$ is required it will probably be quicker to loop the 0 squares.

Example 1.6

Simplify the equation $F = ABCD + \bar{A}BCD + A\bar{C}D + A\bar{C}\bar{D} + \bar{A}B\bar{C}$ and obtain $\bar{F}$.

Solution
The mapping of F is shown.

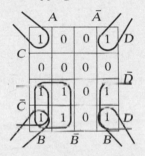

1 Looping the 1 squares, $F = BD + A\bar{C} + B\bar{C}$.
Therefore,

$$\bar{F} = \overline{BD + A\bar{C} + B\bar{C}}$$
$$= \overline{BD}\ \overline{A\bar{C}}\ \overline{B\bar{C}} = (\bar{B} + \bar{D})(\bar{A} + C)(\bar{B} + C)$$
$$= (\bar{A}\bar{B} + \bar{B}C + \bar{A}\bar{D} + C\bar{D})(\bar{B} + C)$$
$$= \bar{A}\bar{B} + \bar{B}C + \bar{A}\bar{B}\bar{D} + \bar{B}C\bar{D} + \bar{A}\bar{B}C + \bar{B}C + \bar{A}C\bar{D} + C\bar{D}$$
$$= \bar{A}\bar{B} + \bar{B}C + C\bar{D} \quad (Ans)$$

2 Looping the 0 squares

$$\bar{F} = \bar{A}\bar{B} + \bar{B}C + C\bar{D} \quad (\text{directly}) \quad (Ans)$$

The Karnaugh map can also be used to simplify an equation of the product-of-sums form. Either the equation can be multiplied out into the equivalent sum-of-product form, or each term can be mapped separately and then the individual maps can be combined by ANDing them. Note that for corresponding squares in each map

$$1 \ 1 = 1 \qquad 1 \ 0 = 0 \qquad 0 \ 0 = 0$$

Example 1.7

Repeat Example 1.3 using a Karnaugh map.

Solution
Applying De Morgan's rule, $F = (A + B)(\bar{A} + \bar{B} + C) + AB$.

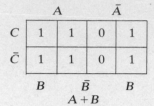

	A		$\bar{A}$	
C	1	1	0	1
$\bar{C}$	1	1	0	1

$B \quad \bar{B} \quad B$
$A + B$

	A		$\bar{A}$	
C	1	1	1	1
$\bar{C}$	0	1	1	1

$B \quad \bar{B} \quad B$
$\bar{A} + \bar{B} + C$

	A		$\bar{A}$	
C	1	1	0	1
$\bar{C}$	0	1	0	1

$B \quad \bar{B} \quad B$
$(A + B)(\bar{A} + \bar{B} + C)$

Lastly, mapping AB

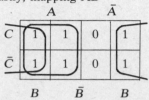

	A		$\bar{A}$	
C	1	1	0	1
$\bar{C}$	1	1	0	1

$B \quad \bar{B} \quad B$

From the mapped squares
$F = A + B \quad (Ans)$

Decimal Representation of Terms

It is sometimes convenient to express the various terms of a Boolean equation in their decimal form. Thus, referring to Table 1.7, assuming A to be the least significant bit, the decimal mapping is as shown.

Decimal mapping

	A		$\bar{A}$		
C	15	13	12	14	D
	7	5	4	6	$\bar{D}$
$\bar{C}$	3	1	0	2	
	11	9	8	10	D

$B \quad \bar{B} \quad B$

Table 1.7

Decimal number	0	1	2	3	4	5	6	7	8	9	10	11	12	13	14	15
A	0	1	0	1	0	1	0	1	0	1	0	1	0	1	0	1
B	0	0	1	1	0	0	1	1	0	0	1	1	0	0	1	1
C	0	0	0	0	1	1	1	1	0	0	0	0	1	1	1	1
D	0	0	0	0	0	0	0	0	1	1	1	1	1	1	1	1

"Don't-Care" Conditions

The truth tables of some logical circuits contain certain combinations of literals for which the output F is unimportant and so they can be either 1 or 0. Such combinations are said to be "don't-care" conditions or states. When a Boolean function is mapped, any don't-care terms are represented by an X and can be looped in with *either* the 1 squares *or* the 0 squares in the simplification of the function. Suppose, for example, that the mapping of a particular function is

	A		Ā		
C	1	1	0	X	D
	1	X	0	1	D̄
C̄	X	1	1	1	
	1	1	0	1	D
	B	B̄	B		

Looping the 1 squares only gives

$$F = AD + ABC + \bar{A}B\bar{D} + \bar{A}B\bar{C} + \bar{B}\bar{C}\bar{D}$$

If the don't-care squares are looped together with the 1 squares, the function can be reduced to the much simpler result:

$$F = A + B + \bar{C}\bar{D}$$

Karnaugh Map for More than Four Variables

The number of squares in a Karnaugh map must be equal to 2^n where n is the number of literals. This means that five literals A, B, C, D and E will require $2^5 = 32$ squares. These can be obtained by drawing two 16-square maps side-by-side, one of which represents E while the other represents $\bar{E}$.

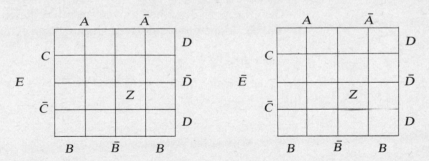

The 32-square map is used to simplify a 5-literal Boolean equation in similar manner to that previously described. Note the similarly positioned squares in each map, e.g. the squares marked Z are considered to be adjacent for looping purposes.

Clearly, the use of the Karnaugh map will become increasingly cumbersome if there are more than five literals and in such cases it is better to employ a tabular method of minimization.

Tabular Simplification of Boolean Equations

The Karnaugh map provides a convenient method for the simplification of Boolean equations with up to four literals and can be used with five literals. When the number of literals is in excess of this number, it is generally easier to employ a tabular method of simplification. The method to be described is due to Quine and McCluskey.

a) Write down the equation to be simplified in its *canonical* form.

b) Write all the canonical terms in a column (column A), starting with those terms that include the *most complemented* literals.

c) In column B write, for each term in *b*), a 1 for an uncomplemented literal and 0 for a complemented literal.

d) Divide column B into *groups* containing, in order, no 1s, a single 1, two 1s, three 1s, and so on.

e) An attempt is now made to "match" each term within a group with another term in another group. Here *matching* means that the two terms differ from one another by only *one* literal. For example, the terms 0001 and 1001 would match. Each term that is matched with another term is ticked. A term should be matched as many times as possible.

f) The *matched terms* are then entered in column C but the two different literals have cancelled out and are hence omitted.

g) When column C has been completed, it is divided into further *groups* as before and another set of matched terms is found.

h) These matched terms are entered in column D and so on.

i) When no more matchings can be found, the table is inspected for all the *unticked* terms. These are known as the PRIME IMPLICANTS. This may be the simplest form of the equations but it is possible that some further simplification can be achieved.

j) A *prime implicant table* is now constructed which is used to locate any common terms and thereby allow a further reduction in the function.

The process as described may seem rather lengthy but with some practice it can be done quite quickly and can be programmed onto a computer.

Example 1.8

Solve the equation $F = AB + B\bar{C} + CD + \bar{B}D$ using a tabular method.

Solution
This is, of course, an equation that could easily be simplified using a Karnaugh map.
Step A Writing the given equation in its canonical form:

$$F = AB\,(C + \bar{C})\,(D + \bar{D}) + B\bar{C}\,(A + \bar{A})\,(D + \bar{D}) + CD\,(A + \bar{A})\,(B + \bar{B})$$

$$+ \bar{B}D\,(A + \bar{A})\,(C + \bar{C})$$

$$= ABCD + AB\bar{C}D + ABC\bar{D} + AB\bar{C}\bar{D} + \bar{A}B\bar{C}D + \bar{A}B\bar{C}\bar{D}$$

$$+ A\bar{B}CD + \bar{A}BCD + \bar{A}\bar{B}CD + A\bar{B}\ddot{C}D + \bar{A}\bar{B}\bar{C}D$$

Table 1.8

	A	B	C	D	E
1	$\bar{A}\bar{B}\bar{C}D$	0 0 0 1 ✓	0 − 0 1 ✓	0 − − 1 ✓	− − − 1 ⎫
2	$\bar{A}B\bar{C}\bar{D}$	0 1 0 0 ✓	0 0 − 1 ✓	− − 0 1 ✓	− − − 1 ⎬ → D
3	$AB\bar{C}\bar{D}$	1 1 0 0 ✓	− 0 0 1 ✓	− 0 − 1 ✓	− − − 1 ⎭
4	$\bar{A}B\bar{C}D$	0 1 0 1 ✓	− 1 0 0 ✓	− 1 0 − ←—————→ $B\bar{C}$	
5	$\bar{A}\bar{B}CD$	0 0 1 1 ✓	0 1 0 − ✓	1 1 − − ←—————→ AB	
6	$A\bar{B}\bar{C}D$	1 0 0 1 ✓	1 1 0 − ✓	− 1 − 1 ✓	
7	$AB\bar{C}D$	1 1 0 1 ✓	1 1 − 0 ✓	− − 1 1 ✓	
8	$ABC\bar{D}$	1 1 1 0 ✓	− 1 0 1 ✓	1 − − 1 ✓	
			0 1 − 1 ✓		
9	$\bar{A}BCD$	0 1 1 1 ✓	0 − 1 1 ✓		
			− 0 1 1 ✓ *Note:* the		
10	$A\bar{B}CD$	1 0 1 1 ✓	1 − 0 1 ✓ same combination		
			1 0 − 1 ✓ is NOT written		
11	$ABCD$	1 1 1 1 ✓	1 1 − 1 ✓ down twice.		
			1 1 1 − ✓		
			− 1 1 1 ✓		
			1 − 1 1 ✓		

Step B These terms are now listed in column A (of Table 1.8).

Step C Column B is then produced by writing 1 for each uncomplemented literal and 0 for each complemented literal in column A.

Step D Column B is then divided into 4 groups by drawing horizontal lines. In each group the number of 1s is the same.

Step E The matching process can now begin. Terms 1 and 4 differ only in their second bit from the left and are matched together and ticked. 0−01 is then entered into column C. Term 1 can also be matched with term 5 and so 00−1 is entered into column C.

Similarly, other matched terms are 1 and 6, 2 and 3, 2 and 4, 3 and 7, 3 and 8, 4 and 7, 4 and 9, 5 and 10, 6 and 7, 6 and 10, 7 and 11, 8 and 11, 9 and 11, and, finally, 10 and 11.

The matching process is now repeated with the entries in column C and so on as shown by Table 1.8.

Two of the column D terms cannot be matched and are therefore *prime implicants*. The terms in column D that can be matched all lead to the same result, i.e. −−−1.

Hence the prime implicants of the function are

$$D + B\bar{C} + AB$$

This may be simplest form possible but a check can be carried out by drawing up a table of the prime implicants. The first step in obtaining the prime implicant table is to write down all the possible combinations for each term. Therefore

$$A\,B\,-\,- \qquad\quad -\,B\,\bar{C}\,- \qquad\quad -\,-\,-\,D$$

1 1 0 0	12	0 1 0 0	4	0 0 0 1	1
1 1 0 1	13	0 1 0 1	5	0 0 1 1	3
1 1 1 0	14	1 1 0 0	12	0 1 0 1	5
1 1 1 1	15	1 1 0 1	13	0 1 1 1	7
				1 0 0 1	9
				1 0 1 1	11
				1 1 0 1	13
				1 1 1 1	15

Hence the prime implicant table is given by Table 1.9.

Table 1.9 Prime implicant table

	1	2	3	4	5	6	7	8	9	10	11	12	13	14	15
AB												√	√	√	√
$B\bar{C}$				√	√							√	√		
D	√		√		√		√		√		√		√		√

It is clear from this table that all three terms must be retained to cover all the required decimal numbers and this means that no further simplification is possible. Hence

$$F = AB + B\bar{C} + D \quad (Ans)$$

The accuracy of this result can easily be checked by means of a Karnaugh map since there are only 4 literals.

The Use of NAND/NOR Gates to Generate AND/OR Functions

The majority of the integrated circuit gates used in modern equipment belong to either the ttl or the cmos logic families. In both of these families NAND and NOR gates are cheaper and faster than the other gates which are available and, in the case of ttl, dissipate less power. It is therefore common practice to construct combinational logic circuits using either NAND or NOR gates *only*.

It is easy to see that the AND function can be obtained by cascading two NAND gates as shown by Fig. 1.6a, and the OR function can be produced by the cascade connection of two NOR gates as in Fig. 1.6b.

Implementation of the AND function using NOR gates, and of the OR function using NAND gates, is not quite as easy to see but the necessary connections are easily deduced with the aid of De Morgan's rules.

Fig. 1.6 Implementation of (a) the AND function using NAND gates, (b) the OR function using NOR gates

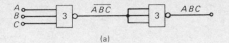 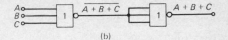

Rule 22 is $\overline{AB} = \bar{A} + \bar{B}$ and hence

$$AB = \overline{\bar{A} + \bar{B}}$$

The right-hand side of this equation is easily implemented using NOR gates as shown in Fig. 1.7a.

The other De Morgan rule (no. 21) is $\overline{A + B} = \bar{A}\bar{B}$ and hence

$$A + B = \overline{\bar{A}\bar{B}}$$

and this can be implemented by NAND gates connected as shown by Fig. 1.7b.

Fig. 1.7 Implementation of (a) the AND function using NOR gates, (b) the OR function using NAND gates

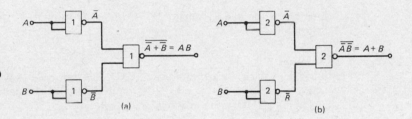

This principle can be extended to three, four, or more literals. For example, the function

$$F = ABCD = \overline{\bar{A} + \bar{B} + \bar{C} + \bar{D}}$$

can be fabricated using the arrangement of Fig. 1.8,

Fig. 1.8 4-input AND function implemented using NOR gates

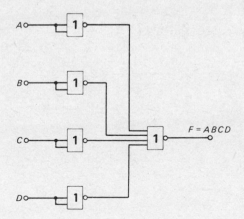

Fig. 1.9(a) Redundant NOR gates, (b) and (c) two ways of implementing

$$F = \overline{A + B + C}$$

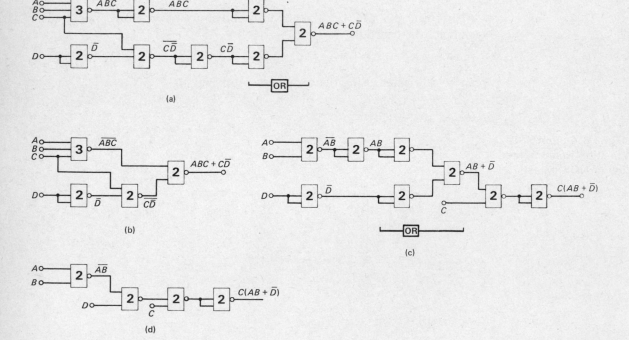

Clearly, more gates are needed to implement the AND/OR functions using NAND/NOR gates but very often the apparent increase in the number of gates required is not as great as may be anticipated. This is because consecutive stages of inversion are redundant (since $\overline{\overline{A}} = A$) and need not be provided (Fig. 1.9a). Also, a circuit of the form given in Fig. 1.9b can be replaced by one 3-input gate as shown by Fig. 1.9c.

In general, Boolean equations in the *product-of-sums* form are best implemented using NOR gates, and *sum-of-products* equations are more easily implemented using NAND gates. This is illustrated by the following example.

Example 1.9

Implement the logical function $F = ABC + C\overline{D}$ using a) NAND gates only, b) NOR gates only.

Solution

First draw the logic diagram using AND, OR and NOT gates and then replace each gate with its a) NAND or b) NOR equivalent circuit. Finally, simplify the circuit if possible by eliminating any redundant gates.

Fig. 1.10

Fig. 1.11

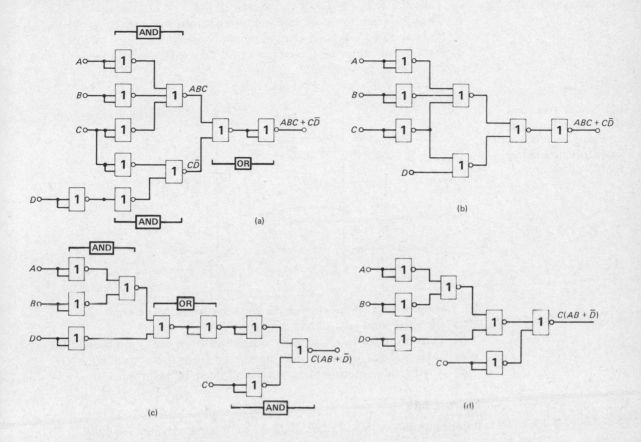

(a)

(b)

(c)

(d)

a) Fig. 1.10*a* shows the given functions implemented using NAND gates only. The circuit includes two sets of redundant gates and these have been removed to produce the circuit of Fig. 1.10*b*.

The equation for F can be written as $F = C\ (\bar{D} + AB)$ and this is implemented by the circuit shown in Fig. 1.10*c* and then further simplified in Fig. 1.10*d*. Although circuits *b* and *d* both use four gates the latter would be the easier to implement since it would only require one quad 2-input NAND gate.

b) The NOR implementation of $F = ABC + C\bar{D}$ is shown in Fig. 1.11*a*. There are only three redundant gates which can be eliminated from the circuit (Fig. 1.11*b*). The alternative is to implement $F = C(AB + \bar{D})$ [see Figs 1.11*c* and *d*].

It is not necessary to follow the procedure used in Example 1.9 since two useful rules are available that allow a NAND or NOR implementation of a function to be written down directly. These two rules are

1 *To implement a sum-of-products equation using NAND gates only*
a) Take the final gate(s) at the output as OR.
b) Take the *even* levels of gate, numbered from the output, as AND.
c) Take the *odd* levels of gate, numbered from the output, as OR.
d) Any literals entering the circuit at an *odd* gate level, numbered from the output, must be inverted.

The rules are illustrated by Fig. 1.12.

Fig. 1.12 Rules for the implementation of a sum-of-products logic function using NAND gates only

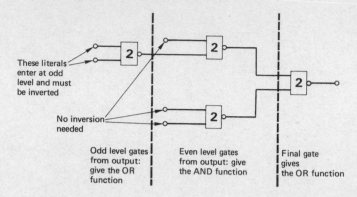

These literals enter at odd level and must be inverted

No inversion needed

Odd level gates from output: give the OR function

Even level gates from output: give the AND function

Final gate gives the OR function

2 *To implement a product-of-sums equation using NOR gates only*
a) Take the final gate at the output as AND.
b) Take the *even* levels of gate, numbered from the output, as OR.
c) Take the *odd* levels of gate, numbered from the output, as AND.
d) Any literals entering at an *odd* level must be inverted.

Example 1.10

Implement using a) NAND gates only, b) NOR gates only, the function

$$F = \bar{A}C + \bar{B}C + A\bar{C}D$$

Solution
a) Using rule **1**, Fig. 1.13a can be drawn directly. To check:

$$F = \overline{(\bar{A}C)\,(\bar{B}C)\,(A\bar{C}D)} = \bar{A}C + \bar{B}C + A\bar{C}D$$

b) The mapping of the function F is

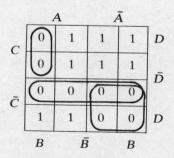

Fig. 1.13

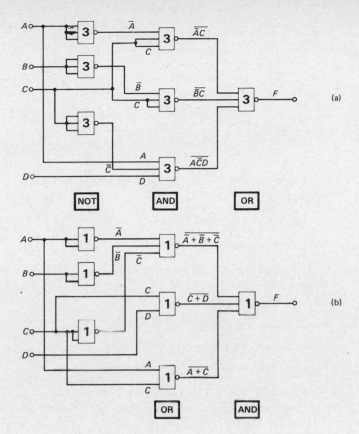

(a)

(b)

From the looped 0 squares $\bar{F} = ABC + \bar{C}\bar{D} + \bar{A}\bar{C}$

Hence $F = \overline{ABC + \bar{C}\bar{D} + \bar{A}\bar{C}}$

$$= (\overline{ABC})\,(\overline{\bar{C}\bar{D}})\,(\overline{\bar{A}\bar{C}})$$

$$= (\bar{A} + \bar{B} + \bar{C})\,(C + D)\,(A + C)$$

Using rule **2**, Fig. 1.13*b* can be drawn. Checking

$$F = \overline{\overline{(\bar{A} + \bar{B} + \bar{C})} + \overline{(C + D)} + \overline{(A + C)}}$$

$$= (\bar{A} + \bar{B} + \bar{C})\,(C + D)\,(A + C)$$

as required.

Multiplying out, $F = \bar{A}C + \bar{B}C + A\bar{C}D$, the original equation.

If the function $F = AB + CD$ is implemented using NAND gates, Fig. 1.14*a* results. If, now, each NAND gate is replaced by a NOR gate (Fig. 1.14*b*), the output F from the circuit is

$$F = \overline{\overline{A + B} + \overline{C + D}} = (A + B)\,(C + D)$$

i.e. the *dual* of $F = AB + CD$. This result provides another way of obtaining the NOR gate implementation of a Boolean equation.

Fig. 1.14 Replacing NAND gates with NOR gates gives the dual of a function

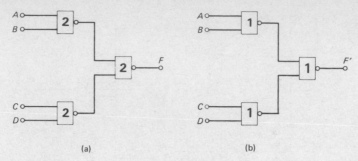

(a)   (b)

Consider once again the function given in Example 1.10. The dual of this equation is

$$F' = (\bar{A} + C)(\bar{B} + C)(A + \bar{C} + D)$$

Multiplying out, $F' = AC + \bar{A}\bar{B}\bar{C} + CD$

This equation can be implemented by NAND gates (Fig. 1.15a) and then each gate can be replaced by a NOR gate (Fig. 1.15b). The output F of the final NOR gate is

$$F = \overline{\overline{A + C} + \overline{C + D} + \overline{\bar{A} + \bar{B} + \bar{C}}}$$

$$= (A + C)(C + D)(\bar{A} + \bar{B} + \bar{C})$$

$$= \bar{A}C + \bar{B}C + A\bar{C}D \quad \text{as before}$$

Figs. 1.13b and 1.15b should be compared.

Another method consists of *doubly* inverting the equation to be implemented. Thus, for the equation $F = \bar{A}C + \bar{B}C + A\bar{C}D$ (again),

$$\bar{F} = \overline{\bar{A}C + \bar{B}C + A\bar{C}D} = (\overline{\bar{A}C})(\overline{\bar{B}C})(\overline{A\bar{C}D}) \quad \text{and} \quad F = \overline{(\overline{\bar{A}C})(\overline{\bar{B}C})(\overline{A\bar{C}D})}.$$

This equation shows immediately that, if the complements of literals A, B and C are available, four NAND gates are required. The implementation is shown by Fig. 1.13a.

Similarly, if $F = (\bar{A} + \bar{B} + \bar{C})(C + D)(A + C)$

$$\bar{F} = (\overline{\bar{A} + \bar{B} + \bar{C}}) + (\overline{C + D}) + (\overline{A + C})$$

$$F = \overline{(\overline{\bar{A} + \bar{B} + \bar{C}}) + (\overline{C + D}) + (\overline{A + C})}.$$

Four NOR gates are necessary (assuming the required complements are available) and the circuit is given by Fig. 1.13b.

Fig. 1.15 NOR implementation of $F = \bar{A}C + \bar{B}C + A\bar{C}D$

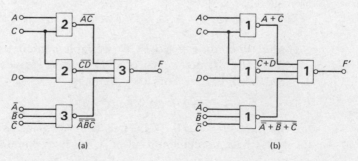

(a)   (b)

Designing a Circuit from a Truth Table

In the design of a combinational logic circuit, the truth table of the required logical operation should be written down and then used to obtain an expression for the output F of the circuit. This expression can then be simplified, using one of the methods presented earlier, before it is implemented by the suitable interconnection of a number of gates. Each 1 appearing in the output column of the truth table must be represented by a term in the Boolean equation describing the circuit. Every such term must contain each literal that is in the logical 1 state and the complement of each literal that is the logical 0 state. More combinational logic circuits will be considered in Chapter 3; here the design method will be illustrated by the design of *half-adder* and *full-adder* circuits.

The **half-adder** is a circuit which adds two inputs A and B to produce a *sum* and a *carry* but which does not take any account of a carry originating from a previous stage. The truth table of a half-adder is given by Table 1.10.

Table 1.10   Half-adder truth table

A	0	1	0	1
B	0	0	1	1
Sum S	0	1	1	0
Carry C	0	0	0	1

Two different equations can be obtained from this truth table to express the sum and the carry C of A and B.

$$\text{Thus}\quad S = A\bar{B} + \bar{A}B \qquad\qquad C = AB$$
$$\text{or}\quad S = (A + B)(\overline{AB}) \qquad C = AB$$

It should be noted that $S = A\bar{B} + \bar{A}B$ is the exclusive-OR function. The implementation of the sum and carry equations using NAND gates is given in Fig. 1.16a. The NOR version of the circuit will be derived using the duality method described earlier.

$$\text{If}\quad S = A\bar{B} + \bar{A}B$$
$$\text{then}\quad S' = (A + \bar{B})(\bar{A} + B) = AB + \bar{A}\bar{B}$$

Fig. 1.16 Half-adder circuits (a) using NAND gates only, (b) using NOR gates only

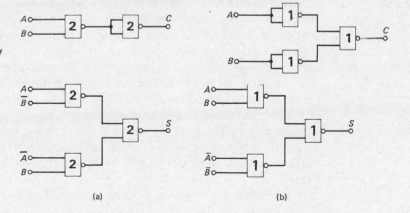

(a)   (b)

The expression for S is easily implemented using NAND gates, and replacing each NAND gate by a NOR gate gives Fig. 1.16b. Checking

$$S = \overline{\overline{A+B} + \overline{\overline{A}+\overline{B}}} = (A+B)(\overline{A}+\overline{B}) = A\overline{B} + \overline{A}B$$

as required. (This method gives a simpler circuit than is obtained by replacing AND/OR gates by their NOR equivalents, see [DT & S].)

Table 1.11 Full-adder truth table

A	0	1	0	0	1	1	0	1
B	0	0	1	0	1	0	1	1
C_{in}	0	0	0	1	0	1	1	1
Sum S	0	1	1	1	0	0	0	1
C_{out}	0	0	0	0	1	1	1	1

The truth table of a **full-adder** is given by Table 1.11; this circuit adds together two inputs A and B and a carry C_{in} from a previous stage to produce a Sum and a Carry C_{out}. From the truth table,

$$S = A\overline{B}\overline{C}_{in} + \overline{A}B\overline{C}_{in} + \overline{A}\overline{B}C_{in} + ABC_{in}$$
$$= (A\overline{B} + \overline{A}B)\overline{C}_{in} + (AB + \overline{A}\overline{B})C_{in}$$
$$= (\text{exclusive-OR}) \ \overline{C}_{in} + (\text{exclusive-NOR}) \ C_{in}$$

Also from the truth table,

$$C_{out} = AB\overline{C}_{in} + A\overline{B}C_{in} + \overline{A}BC_{in} + ABC_{in}$$
$$= AB(\overline{C}_{in} + C_{in}) + (A\overline{B} + \overline{A}B)C_{in}$$
$$= AB + (\text{exclusive-OR}) \ C_{in}$$

The implementation of a full-adder is left as an exercise (see Exercise 1.27). Complete full-adders are also available in the ttl and cmos families, e.g. the 7482 and the 4008.

Race Hazards

A **static hazard** is said to exist in a combinational logic circuit when the change of a single literal from 0 to 1, or from 1 to 0, causes a transient change, or *spike*, at the output when no change should occur.

Consider a circuit which produces an output $F = \overline{A}B + AC$. The use of an inverting stage to obtain $\overline{A}$ means that $\overline{A}$ must always have a small time delay relative to A at the output of the circuit. When $B = C = 1$, then F should be of the form $\overline{A} + A = 1$. Suppose that A changes from 1 to 0. Then $\overline{A}$ takes a short time to change from 0 to 1, and hence for a short while $A + \overline{A}$ will be equal to 0, producing a short, unwanted 0 spike at the output. Similarly, when A changes from 0 to 1, a short period of time will exist during which *both* A and $\overline{A}$ will be a logical 1 but this will not affect the output.

A static hazard exists in a circuit if the literals can alter to produce a change between adjacent cells in the Karnaugh mapping of the function that are *not* looped together. Thus, in the mapping for $\bar{A}B + AC$, a hazard exists since the squares marked in their corner with an X are not looped together. A hazard can be removed by ensuring that such adjacent cells *are* included within a loop. Clearly this will mean the introduction of one, or perhaps more, redundant terms.

If the squares marked X are looped together, the extra term will be BC so that $F = \bar{A}B + AC + BC$. Now, when $B = C = 1$, the output will be 1 regardless of the instantaneous state of A.

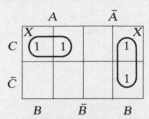

A **dynamic hazard** is said to exist when there are at least three paths, each with a different time delay, for a literal to contribute to the output. Then it is possible for the output to exhibit three or more spikes as the literal changes states.

Logic Design Using Multiplexers

The availability of integrated circuit devices known as multiplexers provides an alternative to the use of gates for the implementation of logic functions. The **multiplexer** or **data selector** is a circuit which will select *one* out of n input lines where n is a power of 2. The design of such circuits using separate logic gates and inverters is considered in Chapter 3 (p. 76). Usually an msi device would be used, several of which are available in both the ttl and cmos families. The block diagram of one-half of a dual 4-line to 1-line multiplexer, the ttl 74LS153, is shown by Fig. 1.17. When the enable input is low, the circuit is able to switch any of four inputs S_0, S_1, S_2 or S_3 to the output terminal. The truth table describing the circuit's logical operation is given by Table 1.12 where X indicates the don't-care condition.

Table 1.12 Multiplexer truth table (4 inputs)

Select Inputs		Data Inputs				Outputs
X	Y	S_0	S_1	S_2	S_3	
0	0	0	X	X	X	0
0	0	1	X	X	X	1
0	1	X	0	X	X	0
0	1	X	1	X	X	1
1	0	X	X	0	X	0
1	0	X	X	1	X	1
1	1	X	X	X	0	0
1	1	X	X	X	1	1

Fig. 1.17 4-to-1 mul-
tiplexer

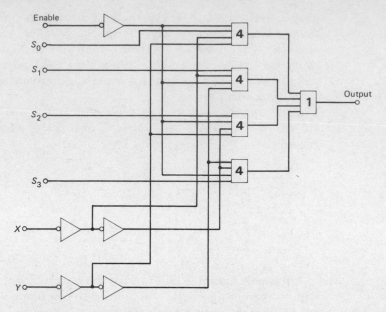

When both the select inputs X and Y are low, the gate connected to S_0 input is the only one that is not inhibited by either X or Y and so the output takes up the S_0 state. Similarly, any other combination of the select input states enables just one of the four AND gates and connects the output terminal to the appropriate input terminal.

From Table 1.12, the Boolean equation describing the operation of the circuit is

$$F = \bar{X}\bar{Y}S_0 + \bar{X}YS_1 + X\bar{Y}S_2 + XYS_3 \tag{1.6}$$

There are four possible values that can be applied to each of the four data input lines; these are 1, 0, C and $\bar{C}$ where C is a literal. There are two literals, A and B, which can be applied to the select input lines. With three literals A, B and C, there are three different ways in which the select variables can be chosen, i.e. AB, AC or BC.

For each possibility a Karnaugh map can be produced which will show the various positions of the data inputs S_0, S_1, S_2 and S_3.

Suppose, firstly, that A and B are chosen to be the select input or *control* variables, then equation (1.6) becomes

$$F = \bar{A}\bar{B}S_0 + \bar{A}BS_1 + A\bar{B}S_2 + ABS_3$$

and the Karnaugh mapping is as shown by Map A.

	A		$\bar{A}$	
C	S_3	S_2	S_0	S_1
$\bar{C}$	S_3	S_2	S_0	S_1
	B	$\bar{B}$		B

Map A

Similarly, if A and C are chosen to be the control variables, equation (1.6) is written as

$$F = \bar{A}\bar{C}S_0 + \bar{A}CS_1 + A\bar{C}S_2 + ACS_3$$

and this is mapped by Map B. Lastly, Map C gives the mapping obtained if B and C are the control variables.

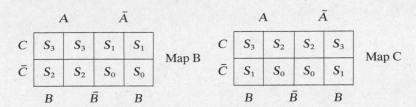

In the design of a multiplexer solution to a logic problem, an arbitrary choice must first be made as to which two of the three literals A, B, and C are to be chosen as the control variables and then the appropriate Map (A, B or C) will be used.

Suppose the Boolean equation

$$F = ABC + \bar{A}\bar{B}C + \bar{A}B\bar{C} + A\bar{B}\bar{C}$$

is to be implemented using a 4-input multiplexer. The mapping of this equation is given by Map D.

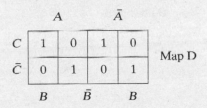

If A and B are chosen to be the control variables this map must be compared with Map A. Hence,

$$S_0 = C + 0 = C = S_3 \qquad S_1 = S_2 = \bar{C}$$

and the required circuit is shown by Fig. 1.18a.

If the control variables are chosen to be A and C, Maps B and D must be compared. Now,

$$S_0 = B = S_3 \qquad S_1 = S_2 = \bar{B} \quad \text{(see Fig. 1.18b)}$$

Lastly, for control variables B and C,

$$S_0 = A \qquad S_1 = \bar{A} \qquad S_2 = \bar{A} \qquad S_3 = A \quad \text{(Fig. 1.18c)}$$

The procedure can be extended to functions with four literals. Suppose the function

$$F = ABCD + \bar{A}\bar{B}CD + A\bar{B}C\bar{D} + AB\bar{C}\bar{D} + A\bar{B}\bar{C}D$$

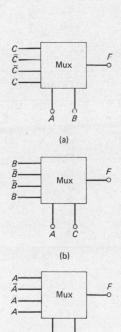

Fig. 1.18 Multiplexer solutions to
$$F = ABC + \bar{A}\bar{B}C$$
$$+ \bar{A}B\bar{C} + A\bar{B}\bar{C}$$

is to be implemented. As before, two of the literals must be arbitrarily chosen to be the control variables; here A and B have been so selected. The 16-square Karnaugh map of the basic multiplexer equation (1.6) is given by Map E (an extension of Map A), while Map F represents the equation to be implemented. Comparing the two tables gives

$$S_0 = CD \qquad S_1 = 0 \qquad S_2 = C\bar{D} + \bar{C}D \qquad S_3 = CD + \bar{C}\bar{D}$$

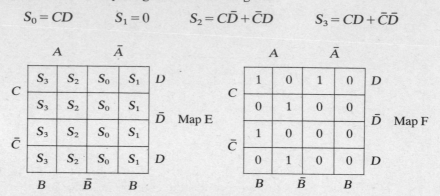

Alternatively, the equation can be implemented using an 8-input multiplexer such as the 14LS151. The truth table of such a device is given by Table 1.13.

Table 1.13 Multiplexer truth table (8 inputs)

Select inputs			Data inputs								Output
X	Y	Z	S_0	S_1	S_2	S_3	S_4	S_5	S_6	S_7	
0	0	0									S_0
1	0	0									S_1
0	1	0									S_2
1	1	0									S_3
0	0	1									S_4
1	0	1									S_5
0	1	1									S_6
1	1	1									S_7

Only the output has been given for each combination of select inputs to simplify the table. The Boolean equation describing this operation is

$$F = S_0\bar{X}\bar{Y}\bar{Z} + S_1 X\bar{Y}\bar{Z} + S_2\bar{X}Y\bar{Z} + S_3 XY\bar{Z}$$
$$+ S_4\bar{X}\bar{Y}Z + S_5 X\bar{Y}Z + S_6\bar{X}YZ + S_7 XYZ \qquad (1.7)$$

Comparing equation (1.7) with the equation to be implemented it can be seen that, letting $A = X$, $B = Y$, and $C = Z$, there is no term in $\bar{A}B\bar{C}$ and so $S_0 = 0$. Similarly, the fifth term of the function, i.e. $AB\bar{C}D$ corresponds with the second term of (1.7), hence $S_1 = D$. Carrying on this way, $S_2 = 0$, $S_3 = \bar{D}$, $S_4 = D$, $S_5 = \bar{D}$, $S_6 = 0$, and $S_7 = D$.

Wired-OR and AND-OR-Invert Logic

Some of the gates in the ttl logic family are designated as *open-collector* types. These are gates that can have their outputs directly paralleled and connected via a *pull-up resistor* to the positive supply voltage (see Fig. 1.19). The outputs of the individual gates are $\overline{AB}$ and $\overline{CD}$; if the output of either gate goes to zero volts the output of the paralleled gates must also become

Fig. 1.19 Direct connection of open-collector gates

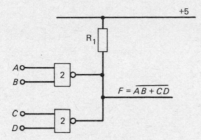

$$F = \overline{AB} + \overline{CD}$$

0 V. Only if both outputs are at logical 1 can the combined output be 1. The logical function performed by a **wired-OR gate** is hence

$$\overline{AB}\,\overline{CD} = \overline{AB + CD}$$

Suppose the function $F = AC + BC$ is to be implemented using wired-OR logic. The mapping of this function is

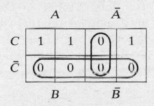

The inputs to the wired-OR gates should be the terms that correspond to the looped 0s in the mapping, i.e. $\bar{C}$ and $\bar{A}\bar{B}$, as shown by Fig. 1.20.

Fig. 1.20 Implementation of $F = AC + BC$ using wired-OR logic

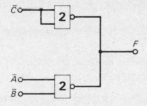

NOR gates do not give a similar result; if open-circuit collector NOR gates have their output terminals paralleled, the effect is to increase their *fan-out* capability.

The **AND-OR-Invert (AOI) gate,** available in both the ttl and cmos logic families, provides the logic circuit shown in Fig. 1.21a. The logical function performed by the AOI gate is

$$F = \overline{AB + CD}$$

Fig. 1.21 (*a*) 2-wide
AOI gate, (*b*) 4-wide
2-input AOI gate

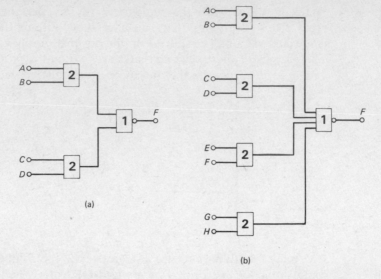

(a)

(b)

Note that this is the same as the function provided by the wired-OR circuit. The circuit is said to be a 2-wide 2-input AOI gate because there are two 2-input AND gates working to the output NOR gate. Similarly, a 4-wide 2-input AOI gate would have four 2-input AND gates (Fig. 1.21*b*). However, some ic gates may have more inputs than their name indicates; for example the ttl dual 2-wide 2-input AOI gate 74LS51 has one gate as per Fig. 1.21*a* but the other gate includes two 3-input AND gates.

Exercises 1

1.1 Explain with the aid of a logic diagram what is meant by an AND-OR-Invert gate. Make clear the meaning of the term *wide* when applied to such a gate. Show how a 2-wide 2-input AOI gate can be used to produce the exclusive-OR function.

1.2 What is meant by duality in Boolean algebra? Find the complements of the equations

(i) $F = C(AB + \bar{A}\bar{B})$
(ii) $F = AC(A + B)(\bar{A} + \bar{B})$

by *a*) the use of duality, *b*) using a Karnaugh map. Implement the complements using either NAND or NOR gates only.

1.3 What are race hazards in a logic circuit? Plot the function

$$F = ABCD + \bar{A}\bar{C}\bar{D} + A\bar{B}D + A\bar{C}D + \bar{A}C\bar{D}$$

on a Karnaugh map. Thence obtain *a*) the minimal solution, *b*) the simplest race-free solution. Implement both *a*) and *b*) using either NAND or NOR gates only.

1.4 The function $F = C(A + \bar{A}B + \bar{A}\bar{B}D)$ is to be implemented using (i) NAND gates, (ii) NOR gates only.
  Simplify the equation and draw suitable circuits. What is meant by wired-OR logic? Could it be used to implement *F*?

1.5 The function $F = CD + \bar{B}\bar{C} + \bar{A}BD$ is to be implemented.

 a) Simplify the equation using a Karnaugh map and implement using NAND gates only.

 b) Repeat *a*) using 2-input NAND gates only.

 c) Obtain $\bar{F}$ from the map and invert to get F. Implement the result using NOR gates only.

 d) Repeat *c*) using 2-input NOR gates only. Comment on your answers to *a*) and *b*), and to *c*) and *d*).

1.6 Use the Quine-McCluskey tabular method to simplify the equation

$$F = \bar{A}\bar{B}CD + A\bar{B}CD + AB\bar{C}D + \bar{A}BCD + \bar{A}B\bar{C}D + A\bar{B}\bar{C}\bar{D}$$
$$+ \bar{A}\bar{B}C\bar{D} + A\bar{B}C\bar{D} + AB\bar{C}\bar{D}$$

1.7 What is meant by the terms: prime implicant, literal, minterm and maxterm?

 Table 1.14 shows a prime implicant chart which has been produced by a Quine-McCluskey reduction of a Boolean equation. Obtain the minimum coverage for the function F.

Table 1.14

	0	1	2	3	4	5	6	7	8	9	10	11	12	13	14	15
$\bar{A}\bar{B}\bar{C}$	√		√											√		
AB		√									√					
BCD	√										√	√				

1.8 Implement the function $F = \overline{(A + B + C)}\,(AB + \bar{A}\bar{B} + AC) + ABC$ using *a*) AOI gates, *b*) NAND gates only, *c*) NOR gates only.

1.9 A circuit is required, using the minimum number of NAND gates only, that has two outputs F_1 and F_2 where

$$F_1 = \bar{B}C\bar{D} + \bar{A}\bar{B}CD + \bar{A}B\bar{C}D$$
$$F_2 = A\bar{B}C\bar{D} + \bar{A}BD + \bar{A}BC\bar{D}$$

Draw the circuit.

1.10 Map the function $\sum(1, 3, 5, 7, 9, 11, 12)$ with don't-care conditions 2, 8 and 10. Obtain two simplified Boolean equations, one suitable for NAND gate, the other for NOR gate implementation. Draw the two logic circuits. Obtain the Boolean equations describing the circuit whose truth table is given by Table 1.15. Simplify the equation and then implement it using (i) NAND gates only, (ii) NOR gates only.

Table 1.15

A	0	0	0	0	0	1	1	1	1
B	0	0	1	1	0	0	0	1	1
C	0	1	0	1	0	1	0	0	1
F	1	1	1	1	0	0	0	0	1

1.11 Implement the function

$$F = ACD + \bar{A}BCD + \bar{B}\bar{C}D + \bar{A}B\bar{C}\bar{D}$$
$$+ \bar{A}\bar{B}CD + A\bar{B}C\bar{D}$$

using (i) a 4-input multiplexer, (ii) an 8-input multiplexer.

Fig 1.22

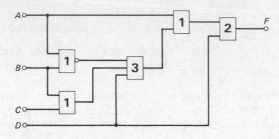

1.12 Determine the Boolean expression for the output of the circuit shown in Fig. 1.22. Simplify the expression if possible and then implement the circuit using (i) NAND gates, (ii) NOR gates only.

Short Exercises

1.13 Use a truth table to show that the logic required for an electric light to be switched on and off from either one of two points is the exclusive-OR function.

1.14 Simplify, using a Karnaugh map,

    a) $F = \bar{A}\bar{B} + \bar{A}B + A\bar{B}$
    b) $AB + \bar{A}BC + \bar{B}C$

1.15 Show that

    a) $A + \bar{A}B = AB$
    b) $AB + \bar{B}C + AB\bar{D} + A\bar{B}C + ABCE = AB + \bar{B}C$
    c) $AC + \bar{B}\bar{C}D + BC\bar{D} + \bar{A}\bar{B}D + \bar{A}\bar{B}\bar{C}D + A\bar{B}\bar{C}\bar{D} = AC + CD + \bar{C}\bar{D} + \bar{C}\bar{B}$

1.16 *a)* Express $F = \overline{\overline{A + BC}} + (A + \bar{C})B$ in sum-of-products form.
    b) Express $F = AB + AC + B\bar{C}$ in product-of-sums form.

1.17 Implement the function $F = A\bar{B}C + \bar{A}B\bar{C}$ using NOR gates only.

1.18 Show how the exclusive-OR function can be produced using open-collector NAND gates and wired-OR logic.

1.19 Implement the function $F = A\bar{B}C + \bar{A}B\bar{C} + A\bar{B}\bar{C} + \bar{A}BC$ using a 4-input multiplexer.

1.20 Plot the function $F = AC\bar{D} + \bar{B}\bar{C}\bar{D} + \bar{B}C\bar{D} + \bar{A}B\bar{C} + \bar{A}CD$ on a Karnaugh map and indicate where static hazards exist. State how the hazards could be eliminated.

1.21 Simplify both algebraically and by mapping each of the following:

    a) $F = AB + AC + BC$
    b) $F = AB\bar{C} + \bar{A}BC + A\bar{B}C\bar{D} + \bar{A}\bar{B}D$
    c) $F = \bar{A}B\bar{C}D + A\bar{B}CD + \bar{A}\bar{B}C\bar{D}$

1.22   Simplify
   a) $A(A+B)B$
   b) $AB(C+\bar{C})$
   c) $A\bar{B}C+A\bar{C}+B$
   d) $\bar{A}+AB+\bar{A}C$
   e) $\bar{A}B\bar{C}+BC$

1.23   Map the expression $F=\sum(3,8,12,14,15)$. Obtain the minimal expressions for a) F, b) $\bar{F}$.

1.24   Simplify $F=\overline{(\overline{\bar{A}C+D})\,(\bar{A}+\bar{C})\,\bar{B}D}$

1.25   Determine the Boolean equation for the output of the circuit given in Fig. 1.21.

1.26   Prove that the circuit given in Fig. 1.4b generates the exclusive-OR function.

1.27   Draw the circuit of a full-adder.

1.28   Implement the function $F=\overline{(AC+BD)}\,(AB)$ using an AOI gate.

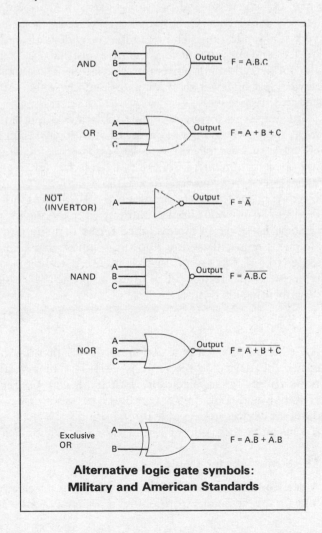

Alternative logic gate symbols:
Military and American Standards

2 Logic Families

The various kinds of gates described in Chapter 1 can be constructed in a number of different ways, some of which use discrete components while others use monolithic integrated circuit techniques. Discrete component logic is nowadays rarely, if ever, employed and it will not be considered in this chapter. The most popular logic families are *transistor-transistor logic* (ttl), of which several versions exist, and *complementary metal oxide semiconductor* (cmos), although *emitter-coupled logic* (ecl) is available for very fast applications. These families provide both *small-scale integrated* (ssi) devices and *medium-scale integrated* (msi) devices, gates being examples of the former and multiplexers of the latter. The terms ssi and msi refer to devices having fewer than ten gates and fewer than 100 gates (or equivalent circuits), respectively.

Integrated circuits containing 1 to 4 gates are often referred to as *random logic* and have been the mainstay of most logic design in the past. In modern circuitry the use of msi and lsi (*large-scale integration*) devices such as multiplexers and microprocessors have led to fewer applications for random logic but it is expected to continue to be used for such purposes as interfacing to, from, and between the lsi devices. In addition, standard ssi devices can provide simple solutions to many digital requirements.

Some lsi devices employ other forms of logic that are not used for the simpler circuits; these are known as pmos, nmos, and integrated injection logic (I^2L). Lastly, some lsi circuits use a technique known as the charge-coupled device (ccd). All of these approaches to digital circuitry will be described in this chapter.

Electronic Switches

An **electronic switch** is a two-state device that has only two stable states, either the device is ON or it is OFF. The binary states 1 and 0 can be represented by an electronic switch; 1 can be represented by the OFF condition and 0 by the ON condition, or vice versa. Although a number of different devices are capable of operating as two-state circuits, usually either a semiconductor diode or a transistor (bipolar or field effect) is used.

The Semiconductor Diode

A **semiconductor diode** is able to operate as an electronic switch because it offers a low resistance to the flow of electric current in one direction and a

Fig. 2.1 (*a*) Generalized diode current/voltage characteristic
(*b*) Equivalent circuit of a diode when conducting
(*c*) Equivalent circuit of a diode when non-conducting

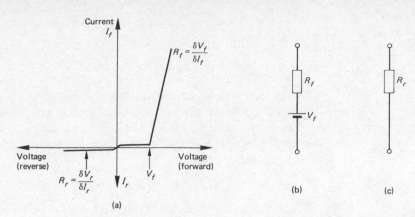

high resistance in the other. When the diode is forward biased and conducts a relatively large current, it is said to be ON. When it is reverse biased and conducts only a very small current, it is said to be OFF. A generalized diode characteristic is given in Fig. 2.1*a*. The diode conducts only when the forward bias voltage is greater than a threshold value labelled as V_f.

When a diode is ON it can be replaced, on paper, by an equivalent circuit consisting of a battery of emf V_f connected in series with a resistance R_f (see Fig. 2.1*b*).

V_f is the voltage dropped across the diode when it is conducting and it has a value that varies between about 0.4 V to 0.75 V depending upon the type of diode, the diode current, and the junction temperature. (A typical temperature coefficient is $-2\text{mV}/^\circ\text{C}$.) For silicon diodes V_f is very often assumed to be 0.6 V.

R_f is the ac resistance of the diode when it is forward biased. When a diode is OFF it can be represented by a resistance R_r (Fig. 2.1*c*), which is the reverse resistance of the diode.

The equivalent circuits of a diode are linear and they assist in the determination of the currents and voltages in a diode circuit.

Example 2.1

Calculate the output voltage of the circuit given in Fig. 2.2 when *a*) $V_1 = +5$ V, $V_2 = 0$ V, *b*) $V_1 = +5$ V, $V_2 = +5$ V, *c*) $V_1 = 0$, $V_2 = +5$ V, and *d*) $V_1 = 0$ V, $V_2 = 0$ V, if $V_f = 0.6$ V and $R_f = 10\,\Omega$.

Solution
a) Diode D_1 will be forward biased and ON while diode D_2 is OFF. Therefore

$$V_{out} = \frac{(5 - 0.6) \times 10\,000}{12\,010} = 3.66 \text{ V} \quad (Ans)$$

b) Both diodes are now ON and therefore

$$V_{out} = \frac{(5 - 0.6) \times 10\,000}{\frac{1}{2} \times 2010 + 10\,000} = 4.0 \text{ V} \quad (Ans)$$

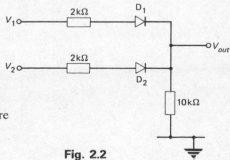

Fig. 2.2

c) This is really the same situation as in *a*). Hence,

$V_{out} = 3.66$ V   (*Ans*)

d) Neither diode conducts and so $V_{out} = 0$ V   (*Ans*)

When the polarity of the voltage applied across a diode is reversed, the diode will not instantaneously change its state from ON to OFF or vice versa. However, the time taken for a diode to switch is generally very small and can be neglected. Suppose a diode has a forward voltage V_F applied to it and is passing a current

$$I_F = (V_F - V_f)/(R_f + R_L) \simeq V_F/R_L$$

If the voltage is suddenly reversed to a new value V_R, the current flowing in the circuit will also reverse its direction and have a magnitude of V_R/R_L for a time, shown in Fig. 2.3 as t_s, during which excess stored minority charge carriers are removed. The time period labelled as t_s is known as the **storage time**. The voltage across the diode does *not* reverse its direction for this period of time. When the excess charges have been removed, the diode voltage reverses its polarity and the diode current starts to fall taking a time t_f to reach zero (Fig. 2.3).

Fig. 2.3 Switching a diode: (*a*) applied voltage, (*b*) diode voltage, (*c*) diode current

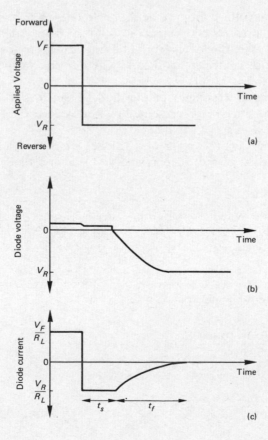

Manufacturer's data generally quote the **reverse recovery time** of a diode. This is the time that elapses from the moment the diode current first reverses its direction of flow to the time the diode current reaches a defined value. The reverse recovery time may vary from a few nanoseconds to a few microseconds.

The Bipolar Transistor

Fig. 2.4 shows a typical set of output characteristics for a **bipolar transistor** with a dc load line drawn between the points

$$V_{CE} = V_{cc} = 12\text{ V}, \ I_C = 0 \quad \text{and} \quad V_{CE} = 0\text{ V}, \ I_C = V_{cc}/R_L = 6\text{ mA}$$

where R_L is the total load on the transistor. When a transistor is used as switch it is rapidly switched between two stable states OFF and ON.

Fig. 2.4 The transistor as a switch

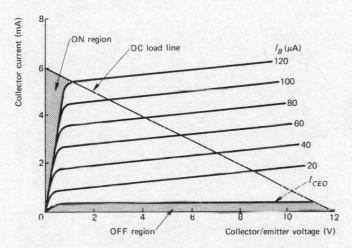

When the transistor is OFF, both its collector/base and emitter/base junctions are reverse-biased and the collector current is only the small collector leakage current I_{CEO}. The collector/emitter voltage V_{CE} of the transistor is then equal to the supply voltage V_{CC}.

When the transistor is conducting current in its active region, the collector/base junction is reverse biased but the emitter/base junction is forward biased. As the base/emitter voltage V_{BE} is increased, the base current increases also and this produces an increase in the collector current since

$$I_C = h_{FE}I_B + I_{CEO} \simeq h_{FE}I_B$$

Eventually the point is reached at which the collector/base junction becomes forward biased and the transistor is said to be **saturated** or **bottomed**. The collector current now has its maximum, or ON, value $I_{C(sat)}$ and the base current is $I_B = I_{C(sat)}/h_{FE}$. The base/emitter voltage $V_{BE(sat)}$ producing this base current usually has a value of about 0.75 V. Any further increase in the base current will *not* produce a corresponding increase in the collector current. The collector/emitter voltage $V_{CE(sat)}$ of the transistor is then very low, being typically in the region of 0.1–0.2 V, because most of the supply voltage is dropped across the collector resistor R_L. Thus, the saturated collector current is equal to

$$I_{c(sat)} = \frac{V_{cc} - V_{CE(sat)}}{R_L} \simeq \frac{V_{CC}}{R_L}$$

(2.1)

Note that $V_{cc}/R_L > h_{FE}I_B$.

Example 2.2

The transistor shown in Fig. 2.5 has $h_{FE} = 50$. Determine whether or not the transistor saturates. If it does not find the maximum base resistance for saturation to occur. $V_{BE} = 0.6$ V and $V_{CE(sat)} = 0.2$ V.

Fig. 2.5

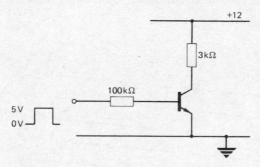

Solution

$$I_B = \frac{5 - 0.6}{100 \times 10^3} = 44 \ \mu A$$

and so $I_C = 50 \times 44 \times 10^{-6} = 0.22$ mA.

When the transistor is saturated

$$I_{C(sat)} = \frac{12 - 0.2}{3 \times 10^3} = 3.93 \text{ mA}$$

For saturation to occur

$$I_B \geqslant \frac{3.93 \times 10^{-3}}{50} \geqslant 78.7 \ \mu A$$

and so the transistor does *not* saturate.

For saturation to take place the maximum base resistance is equal to

$$(5 - 0.6)/78.7 \times 10^{-6} \simeq 56 \text{ k}\Omega \quad (Ans)$$

A transistor can be rapidly switched ON and OFF by the application of a rectangular waveform of sufficient amplitude to its base. In either of the two stable states the power dissipated within the transistor is small because *either* V_{CE} *or* I_C is approximately equal to zero. The active (or amplifying) region of the transistor is rapidly passed through as the transistor switches from one state to the other, and so little power is dissipated.

A transistor is unable to change state instantaneously when the voltage applied to its base is changed, because of charges stored in *a*) the base region, *b*) the collector/base depletion layer, and *c*) the base/emitter depletion layer. When the transistor is OFF both its base/emitter and its collector/base junctions are reverse biased and the two depletion layers are wide. When a voltage is applied to the base to turn the transistor ON, the base current supplies charge to both the p-n junctions and this reduces the widths of the two depletion layers. At some point the base/emitter depletion layer will be sufficiently narrow to allow charge carriers to move from the emitter into the base, and when these reach the collector a collector current starts to flow. As the collector current increases, the collector/emitter voltage falls and the width of the collector/base depletion layer decreases. If the base current is sufficiently large the collector current continues to increase until the transistor is saturated; the collector/base junction is then forward biased and so charge carriers are no longer swept into the collector region. This means that an *excess charge* is stored in the base region.

When a voltage is applied to the base to turn the transistor OFF, the collector current will not start to fall until all the excess base charge has been removed. This time delay is known as the **storage delay**.

Fig. 2.6 Switching a bipolar transistor circuit: (*a*) input voltage, (*b*) output voltage

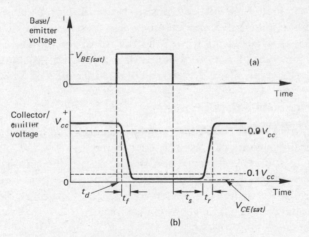

Fig. 2.6 shows how the collector/emitter voltage of an initially OFF bipolar transistor varies when a voltage pulse is applied to its base. The terms shown in the figure are defined thus:

t_d is the time that elapses between the application of the base signal and the collector/emitter voltage falling to 90% of its original value of V_{cc} volts.

t_f is the time taken for the collector/emitter voltage to fall from 90% to 10% of V_{cc} volts.

t_s is the time delay that occurs between the removal of the base voltage and the collector/emitter voltage rising to 10% of its final value of V_{cc} volts.

t_r is the time taken for the collector/emitter voltage to rise from $0.1V_{cc}$ to $0.9V_{cc}$ volts.

Typically, the ON and OFF times are about 6 ns and 10 ns respectively, and to increase the switching speed the transistor must be prevented from

Fig. 2.7 Use of a diode to increase switching speed

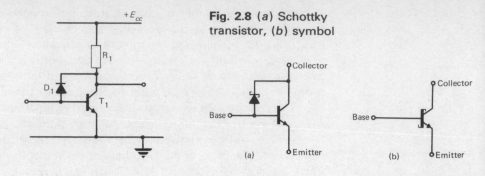

Fig. 2.8 (*a*) Schottky transistor, (*b*) symbol

saturating. This can be achieved by the connection of a diode between the base and the collector terminals of the transistor as shown in Fig. 2.7. When the transistor is turned ON, its collector/emitter voltage falls, and when the collector potential becomes less positive than the base potential the diode D_1 conducts and prevents an excess current from entering the base. As a result the transistor is not driven into saturation and there is no storage of charge in the base region. The best results are obtained if a *Schottky diode* is employed since these devices have zero charge storage and so are very fast switches. A Schottky transistor is a bipolar transistor that has a Schottky diode internally connected between its base and collector terminals, (see Fig. 2.8*a*). The symbol for a Schottky transistor is given in Fig. 2.8*b*. The voltage drop across a Schottky diode, or p-n junction, is smaller than for an ordinary diode, typically 0.3 V when conducting and 0.5 V when full ON.

The Field Effect Transistor

A **field effect transistor** can also be employed as an electronic switch since its drain current can be turned ON and OFF by the application of a suitable gate-source voltage. When the fet is ON, the gate-source voltage will have moved the operating point to the top of the load line (similar to the bipolar transistor switch), and the voltage across the fet, known as the saturation voltage $V_{DS(sat)}$, is very small, typically 0.2 V–1.0 V.

Fig. 2.9 Enhancement mosfet switch; T_1 acts as an active load for T_2

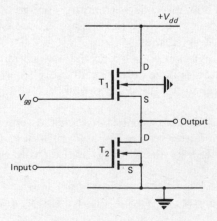

When OFF the fet passes a very small current, typically 1 nA for a junction fet and 50 pA for a mosfet. The drain load resistance needed for a mosfet switch is often of the order of some tens of kilohms and such high values are not conveniently fabricated within a monolithic integrated circuit. For this reason the drain load is often provided by another mosfet as shown in Fig. 2.9. The bottom transistor T_2 is operated as the switch while the upper transistor T_1 is biased to act as an *active load* by the voltage V_{gg}, where $V_{gg} > V_{dd}$. Very often $V_{gg} = V_{dd}$ and a separate bias voltage supply is then not necessary. The active load transistor T_1 is always conducting current.

The switching speed of a fet is determined by the stray and transistor capacitances which are unavoidably present in the circuit. The effects encountered with the bipolar transistor circuit are now insignificant because in a fet current is carried by the majority charge carriers.

Parameters of the Logic Families

The various logic families that will be discussed in this chapter possess different characteristics, which means that any one of them may be the best suited for a particular application. For example, for one application the most important consideration might be the highest possible speed of operation, whilst for another application it might be the minimum possible power dissipation. The characteristics of the various logic families can be classified under the following headings: speed of operation, fan-in and fan-out, noise margin or noise immunity, and power dissipation.

1 Speed of Operation The speed of operation, or the propagation delay, of a logic circuit is the time that elapses between the application of an input signal and the resulting change in the logic state at the output (see Fig. 2.10). Generally, the propagation delay is measured at the 50% points on the input and output waveforms. The two delays shown are not necessarily equal to one another and it is customary to quote their average value.

Fig. 2.10 Speed of operation of a logic device

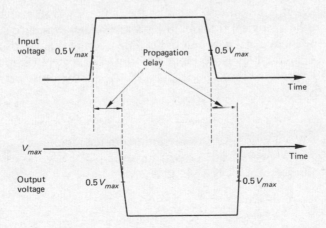

2 Noise Margin or Noise Immunity The false operation of a logic circuit can be caused by transient voltages produced by switching, and by noise voltages induced from other logic lines and/or generated within the power supplies. If the noise voltage is of sufficiently high value, it may cause a gate to change its output state even though the input signal voltage has remained constant. The noise margin or noise immunity of a gate is the maximum noise voltage that can appear at its input terminals without causing a change in the output state.

Fig. 2.11 Noise margin

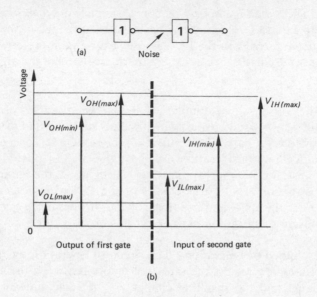

Consider as an example the case of two cascaded NAND or NOR gates shown in Fig. 2.11a. Suppose that the gates have a range of possible high-level and low-level input voltages as shown in Fig. 2.11b. The output voltage of a gate may vary between limits $V_{OH(max)}$ and $V_{OH(min)}$ for the logical 1 output and between the limits $V_{OL(max)}$ and $V_{OL(min)}$ for the output to be at logical 0 (assuming positive logic). Similarly, the input voltage of a gate may vary between limits $V_{IH(max)}$ and $V_{IH(min)}$ for the input state to be at logical 1. For the input state to be at logical 0 the input voltage must be between the limits $V_{IL(max)}$ and $V_{IL(min)}$. The noise margin of a gate will then be

$$V_{OH(min)} - V_{IH(min)} \quad \text{and} \quad V_{IL(max)} - V_{OL(max)}$$

When noise margins are quoted in manufacturer's data sheets it is usual to give the worst-case values; for example the worst-case values for the logic voltages of a ttl NAND gate are

$$V_{OH(min)} = 2.4 \text{ V} \qquad V_{OL(max)} = 0.4 \text{ V}$$
$$V_{IH(min)} = 2.0 \text{ V} \qquad V_{IL(max)} = 0.8 \text{ V}$$

so that the worst-case noise margins are

$$2.4 - 2.0 = 0.4\,\text{V} \quad \text{and} \quad 0.8 - 0.4 = 0.4\,\text{V}$$

This is the guaranteed noise margin for ttl devices but it is typically at least 1 V.

3 Power Dissipation Power is dissipated within a transistor as it switches from one state to another and also within all current-carrying resistors. The d.c. power dissipation of a gate is the product of the d.c. supply voltage and the mean current taken from that supply. Typical figures are quoted in data sheets.

4 Fan-in and Fan-out The *fan-in* of a gate is the number of inputs connected to the gate. The *fan-out* of a gate is the maximum number of similar circuits that can be connected to its output terminals without the output voltage falling outside the limits at which the logic levels 0 and 1 are specified.

Current Sinking and Current Sourcing

Current sinking and current sourcing are two terms commonly employed in digital work to describe the basic operation of a logic circuit.

A **current sink** is a circuit that is supplied with a current by another circuit, while a **current source** supplies a current to a sink load. The difference in meaning between the two terms is illustrated by Fig. 2.12. In Fig. 2.12a current will flow *into* the sink circuit when the switch is closed, while in Fig. 2.12b closure of the switch makes a current flow *out* of the source circuit.

For any logic circuit a sink load is one that goes to earth via a forward-biased p-n junction (diode or the base/emitter junction of a transistor) and tends to reduce the logical 1 level of the circuit driving it (see Fig. 2.12c). Conversely, a source load is one that goes to V_{cc} volts via a forward-biased junction (Fig. 2.12d) and tends to increase the logical 0 output voltage level of the previous circuit.

Sometimes a circuit can be *either* a sink *or* a source depending upon its operating conditions. Consider Fig. 2.13 (which will later be recognized as a

Fig. 2.12 Current sinking and sourcing, (a) and (c) are current sinks, (b) and (d) are current sources.

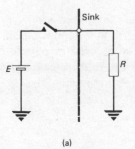
(a)

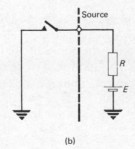
(b)

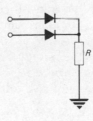
(c)

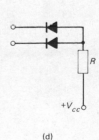
(d)

Fig. 2.13 A circuit that can be either a current sink or a current source

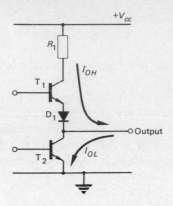

totem-pole output stage for a ttl gate). When T_1 is OFF and T_2 is ON, the output terminal is connected to earth via T_2. This allows a positive supply voltage connected via a forward-biased p-n junction to the output terminal to pass a current I_{OL} via T_2 to earth. Hence T_2 acts as a current sink. Conversely, if T_1 is ON and T_2 is OFF, the output terminal is connected to the positive supply line $+V_{cc}$ via T_1. If, therefore, the output terminal is connected, via a forward-biased p-n junction, to earth, a current will flow from V_{cc} via T_1 to earth so that T_1 acts as a current source.

Logic Levels

Ideally, the logical 0 and 1 voltage levels are 0 V and V_{cc} volts respectively but, in practice, both these levels will differ from the ideal. Consider Fig. 2.14 which shows a transistor that represents the output stage of a gate.

Fig. 2.14 Logic levels

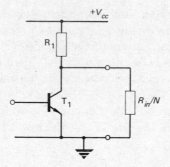

When the transistor is ON the voltage across it—the logic 0 voltage level—is $V_{CE(sat)}$ or approximately 0.2 V. The logical 1 voltage level will depend upon the number N of similar gates (the fan-out) that are connected in parallel across the output terminals. The total resistance across the output terminals is R_{in}/N where R_{in} is the input resistance of each of the N identical gates.

The output 1 voltage level is

$$\frac{V_{cc}R_{in}/N}{R_1 + R_{in}/N}$$

and it is clear that the greater the fan-out, the lower the logical 1 voltage level. The maximum fan-out is limited by the allowable minimum voltage specified for the logical 1 state; for ttl, for example, this is 2.2 V.

Diode-Transistor Logic

The basic gate in the **diode-transistor logic** or **dtl** family performs the NAND function and is shown by Fig. 2.15. When both inputs A and B have a positive voltage of 5 V applied, i.e. both A and B are at logical 1, neither diode conducts and T_1 will be driven into conduction (but not saturation) by the current supplied by R_1. Its emitter current flows in R_3 and develops a voltage that turns T_2 ON. Thus the output voltage of the gate is approximately zero volts and the NAND function has been performed.

Fig. 2.15 DTL NAND gate

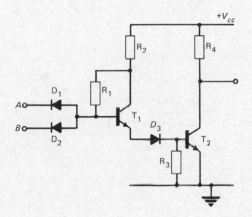

When any one, or more, of the inputs is low (binary 0) the associated diode conducts and a current flows from the gate into the driving circuit (which is therefore a sink). The base voltage of T_1 is then equal to the logical 0 output voltage level of the preceding stage (approximately 0.2 V) plus the voltage drop across the diode (0.7 V). This voltage is low and so the other diodes remain OFF, and T_1 turns OFF. The emitter current of T_1 is zero and so T_2 turns OFF also and the output voltage rises to $+V_{cc}$ volts, i.e. logical 1.

Example 2.3

The dtl NAND gate shown in Fig. 2.15 has the following values: $R_1 = 2$ kΩ, $R_2 = 1800$ Ω, $R_3 = 5$ kΩ and $R_4 = 2$ kΩ. The transistors have a minimum h_{FE} of 50, $V_{BE} = 0.6$ V, $V_{BE(sat)} = 0.7$ V and $V_{CE(sat)} = 0.2$ V. The ON voltage of a diode is 0.7 V and the collector supply voltage $V_{cc} = +5$ V. Calculate the currents and voltages in the circuit when the output terminal is connected to the input of *one* similar gate.

Solution
When one of the inputs is at logical 0, or 0.2 V, the associated diode conducts and the base voltage of T_1 is $0.2 + 0.7 = 0.9$ V. For T_1 to conduct, its base voltage must be equal to $0.7 + 0.6 = 1.3$ V and so T_1 is OFF. Diode D_3 is also OFF. The base/emitter voltage of T_2 is then 0 V, and hence T_2 is also OFF. The input diode to the next gate will also be OFF and so the output voltage rises to +5 V.

When all of the inputs are at +5 V the input diodes D_1 and D_2 turn OFF. The base voltage of T_1 is then

$$V_{BE2(sat)} + V_{BE1} + V_{D3} = 0.6 + 0.7 + 0.7 = 2.1 \text{ V}$$

and T_1 conducts. Then $I_{C1} = 50 I_{B1}$

Therefore,

$$5 - 2.1 = 1.8 \times 10^3 \times 51 I_{B1} + 2 \times 10^3 I_{BI}$$

and so

$$I_{B1} = 2.9/93.8 \times 10^3 \simeq 31 \ \mu A$$
$$I_{C1} = 50 \times 31 \ \mu A = 1550 \ \mu A = 1.55 \text{ mA}$$

The current flowing in R_3 is $0.7/5 \times 10^3 = 0.14$ mA and so

$$I_{B2} = 1.58 - 0.14 = 1.44 \text{ mA}$$

With *no* external load connected

$$I_{C2} = (5 - 0.2)/2 \times 10^3 = 2.4 \text{ mA}$$

When the fan-out is 1 the current sourced by the load is

$$(5 - 0.7 - 0.2)/(2 \times 10^3 + 1.8 \times 10^3) \simeq 1.08 \text{ mA} \quad (Ans)$$

The calculation of the maximum fan-out of this circuit is left as an exercise (Exercise 2.13).

Transistor-Transistor Logic

The most popular and widely used logic family is the **transistor-transistor logic** or **ttl** family. TTL is manufactured in integrated circuit form by several manufacturers. The great popularity of this logic family arises because it offers fairly high speed, particularly the Schottky versions, good fan-in and fan-out, and it is easily interfaced with other digital circuitry. In addition, it is cheap and readily available from many sources.

The standard ttl logic, known as the 54/74 series, has a poor noise immunity and a rather high power consumption. The 74 series is designed for commercial applications and can operate at ambient temperatures of up to 70°C. The 54 series is primarily intended for military applications and has a maximum ambient temperature figure of 125°C.

The circuit of a standard **ttl NAND gate** is shown in Fig. 2.16. Only two inputs are shown but the fan-in may be up to 8. The *totem pole* output stage provided by transistors T_3 and T_4 is used since it provides a low output impedance and the capability to both sink and source currents.

When both input terminals are at logical 1 ($\simeq +5$ V) the emitter/base junctions of the multiple-emitter transistor T_1 are reverse biased *but* its collector/base junction is forward biased. This means that the operation of T_1 is *inverted* and its current gain h_{FE1} is less than unity. Current flows from the collector power supply through R_1 and T_1 into the base of transistor T_2. The base current of T_2 is $-(1 + h_{FE1})I_{B1}$ which is large enough to drive T_2 into saturation and so the collector/emitter voltage of T_2 is approximately 0.2 V.

Fig. 2.16 Standard ttl
NAND gate

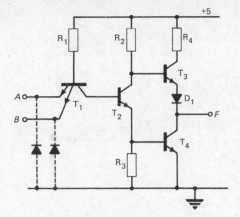

The base/emitter voltage V_{BE4} of T_4 is the voltage developed across R_3 and this turns T_4 ON so that the output voltage falls to $V_{CE4(sat)}$ or about 0.2 V. The base/emitter voltage of T_3 is now equal to

$$V_{CE2(sat)} + V_{BE4(sat)} - (V_{CE4(sat)} + V_{D1})$$
$$-0.2 + 0.75 - (0.2 + 0.6) = 0.95\ \text{V} - 0.8\ \text{V} = 0.15\ \text{V}$$

and so this transistor turns OFF. Thus, the logical state of the output is 0 when both the inputs are at logical 1 and so the NAND function is performed. If the diode D_1 were not present the base/emitter voltage V_{BE3} of T_3 would be equal to $0.95 - 0.2 = 0.75$ V and this would be large enough for T_3 to conduct. With D_1 in circuit

$$V_{BE3} = 0.95 - 0.6 - 0.2 = 0.15\ \text{V}$$

When one or more of the inputs is at logical 0 (approximately 0.2 V), the associated emitter/base junction(s) of T_1 is forward biased and T_1 is turned fully ON. The collector potential of T_1 is then

$$V_{CE1(sat)} + 0.2 = 0.2 + 0.2 = 0.4\ \text{V}$$

This potential is not large enough to keep T_2 conducting and so T_2 turns OFF. The collector potential of T_2 is now +5 V and its emitter potential is 0 V and T_3 turns ON and T_4 turns OFF. The output voltage does not immediately change its value because of the stray capacitance across the output terminals. T_3 sources current to this capacitance and the output voltage rises exponentially towards +5 V. As the charging current falls, T_4 comes out of saturation and V_{out} attains a steady value equal to $V_{cc} - V_{BE3} - V_{D1}$.

Example 2.4

The circuit of Fig. 2.16 has $R_1 = 4\ \text{k}\Omega$, $R_2 = 1.6\ \text{k}\Omega$, $R_3 = 1\ \text{k}\Omega$, $R_4 = 130\ \Omega$, $h_{FE1} = 1$ and $h_{FE2} = h_{FE3} = h_{FE4} = 30$. Calculate the currents and voltages in the circuit.

Solution
When all the inputs are at logical 1,

$$I_{C1} = (1 + h_{FE1})I_{B1} = 2I_{B1} = I_{B2}$$

T_2 and T_4 are saturated and T_3 is OFF because its base voltage is $0.2 + 0.75 = 0.95$ V, while its emitter voltage is $0.2 + 0.6 = 0.8$ V. The base potential of T_1 is

$$V_{BE4(sat)} + V_{BE2(sat)} + V_{CB1} = 0.7 + 0.7 + 0.7 = 2.1 \text{ V}$$

and hence

$$I_{B1} = (5 - 2.1)/4 \times 10^3 = 0.725 \text{ mA}$$

Also $\quad I_{B2} = 2I_{B1} = 1.45$ mA

The collector current of T_2 is

$$V_{cc} - V_{BE3} = (5 - 0.95) \times 1.6 \times 10^3 = 2.5 \text{ mA}$$

The current flowing in R_3 is $0.7/10^3 = 0.7$ mA.
The emitter current of T_2 is

$$I_{E2} = I_{B2} + I_{C2} = 1.45 + 2.5 = 3.95 \text{ mA}$$

Hence $\quad I_{B4} = 3.95 - 0.7 = 3.25$ mA
T_3 is OFF, hence $I_{C3} = 0$. I_{C4} depends upon the load connected across the output terminals, i.e. upon the fan-out.

If any one or more of the inputs is at logical 0, or approximately 0.2 V, then T_2 and T_4 turn OFF and T_3 saturates. The base potential of T_3 is initially at

$$V_{BE3(sat)} + V_{D1} + V_{out} = 0.75 + 0.7 + 0.2 = 1.65 \text{ V}$$

and so

$$I_{B3} = (5 - 1.65)/1.6 \times 10^3 \simeq 2.1 \text{ mA}$$

Also

$$I_{C3} = (V_{cc} - V_{CE3(sat)} - V_{D1} - V_{out})/R_4$$
$$= (5 - 0.2 - 0.7 - 0.2)/130 = 30 \text{ mA}$$

Note that this means that h_{FE3} must be at least equal to $30/2.1 = 14.3$.

The output voltage rises exponentially towards 5 V and settles down at a steady value of

$$5 - V_{BE3} - V_{D1} = 5 - 0.6 - 0.6 = 3.8 \text{ V}$$

Fan-out

The **fan-out** is limited by the current that T_4 can sink when it is saturated. Referring to Fig. 2.17 with one gate only connected to the output terminals,

$$I_{out} = (V_{cc} - V_{BE1} - V_{out})/R_1 = (5 - 0.75 - 0.2)/4000 = 1.01 \text{ mA}$$

For a fan-out of N the output current will be $1.01N$ mA. For T_4 to be in saturation, $I_{B4}h_{FE4} > 1.01N$ and so the maximum fan-out is given by

$$I_{B4}h_{FE4}/1.01$$

There is, however, another factor to be considered and that is the maximum output voltage that can represent logic 0. The standard value for ttl is 0.4 V and, to keep V_{out} below this, the fan-out is normally restricted to ten.

Fig. 2.17 Calculation
of fan-out

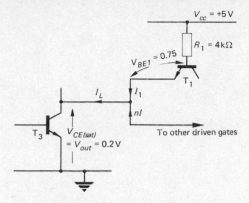

Noise Margins

When all inputs are at logical 1 or $+3.8\,$V, the base potential of T_1 is $0.7+0.7+0.7=2.1\,$V. For a diode to remain in its reverse-biased state when a noise voltage V_n is present, then

$$2.1-3.8-V_n \leqslant 0.7 \quad \text{or} \quad V_n \geqslant 2.1-3.8-0.7 \geqslant -2.4\,\text{V}$$

Similarly, when any one of the inputs is at logical 0, the base potential of T_1 is $0.2+0.2=0.4\,$V. A noise voltage V_n will cause false operation of the circuit if

$$0.4+V_n \geqslant V_{BE2}+V_{BE4} \geqslant 0.7+0.7 \geqslant 1.4\,\text{V}$$

Hence the noise margin is $1.4-0.4=1.0\,$V.

The *worst-case* noise margins are equal at $400\,$mV.

Voltage Levels

For a ttl NAND gate, the input and output voltages required to specify the 1 and 0 logical states are

$$V_{01(min)}=4.4\,\text{V} \qquad V_{00(max)}=0.4\,\text{V}$$
$$V_{11(min)}=2.0\,\text{V} \qquad V_{10(max)}=0.8\,\text{V}$$

The sink current $I_{OL}=16\,$mA and the source current $I_{OH}=0.8\,$mA. These figures can be expressed by the *transfer characteristic* of the gate shown in Fig. 2.18.

The circuits of other gates in the ttl family are shown in [DT&S].

The **threshold voltage** is the input voltage at which a change of the output state of the gate is just triggered. A reasonable approximation to this value is the voltage midway between the $V_{10(max)}$ and $V_{11(min)}$ values, i.e. $1.4\,$V.

Low-power TTL

The series 54/74 L uses the same circuitry as the standard series but all the resistance values are increased in order to minimize the internal power

Fig. 2.18 TTL transfer characteristic

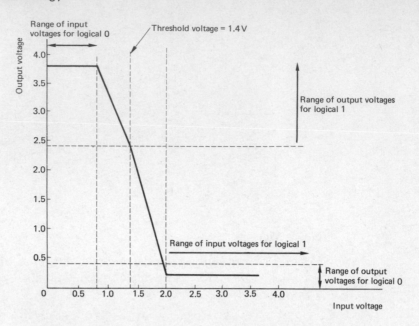

dissipation. This is, however, gained at the expense of a considerable increase in the propagation delay of each circuit.

Schottky TTL

A decrease in the propagation delay of a gate can be achieved if Schottky diodes and transistors are used and a Schottky ttl logic series, 54S/74S, is available from several sources. The circuit of the NAND gate is given in Fig. 2.19 and it can be seen to be similar to the ttl NAND gate except that a) resistor R_3 is replaced by transistor T_4 and R_3/R_5 and b) diode D_1 is replaced by T_3/T_5 to increase the speed of operation.

Fig. 2.19 Schottky ttl NAND gate

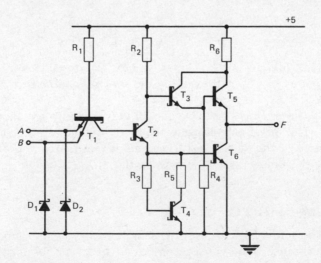

Low-power Schottky TTL

Low-power Schottky ttl, 54LS/74LS, combines the advantages of fast operation, low-power dissipation, and high-frequency capability. A recent introduction to the market is Advanced Low-power Schottky which provides a further four-fold reduction in the speed/power product. This makes it suitable for very fast applications which have hitherto been the province of *emitter-coupled logic* (ecl). Most new designs use the 54/74LS series and it has been forecast that the other versions of ttl will soon become obsolete; indeed some manufacturers have now ceased to make any ttl devices other than LS.

Fig. 2.20 Low-power Schottky ttl NAND gate

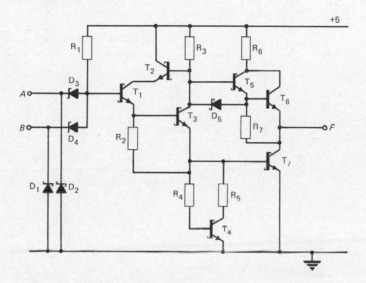

The circuit of the **low-power Schottky NAND gate** is shown in Fig. 2.20. When both inputs are at logical 1, diodes D_1 and D_2 are OFF. T_1 and T_3 turn ON and then the voltage at the base of T_1 is equal to $0.5 + 0.5 + 0.5 = 1.5$ V. The collector potential of T_3 is low and so the Darlington pair T_5/T_6 [EIV] turns OFF. When the base/emitter voltage of T_3 is very nearly equal to $V_{BE(sat)}$, then T_4 will be conducting and will take most of the emitter current of T_3. Transistor T_7 will not receive sufficient base current to turn it ON until the base potential of T_1 is high enough for T_1 to supply sufficient current. This has the effect of squaring the shape of the transfer characteristic and thereby increasing the noise immunity of the circuit (Fig. 2.21).

The speed of switching is increased by T_2 and D_5 whereby T_2 supplies a current surge during the turn-on of T_1 while D_5 provides a discharge path for T_6.

When one or more inputs are at logical 0, the associated input diode conducts and the base potential of T_1 falls below the value needed to keep T_1 conducting. Therefore T_1 turns OFF. This makes transistors T_3 and T_7 turn

Fig. 2.21 Transfer characteristics of ttl and low-power Schottky ttl gates

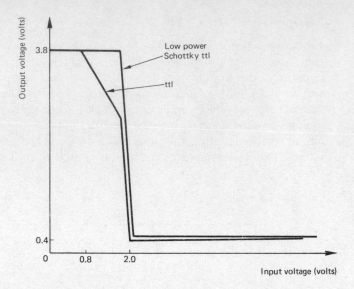

Table 2.1 TTL gates

Type of gate	Standard	low-power	Schottky	Low-power Schottky	Advanced low-power Schottky
Propagation delay (ns)	9	30	3	7	5
Guaranteed noise margin (V)	1 0.4 0 0.4	0.4 0.4	0.7 0.3	0.7 0.3	0.7 0.4
Power dissipation (mW)	10	1	20	2	1
Fan-out	10	10	10	10	10

OFF as well. The collector potential of T_3 rises and turns T_5/T_6 ON. The base of the output transistor T_6 is returned via R_7 to the output terminal since this allows the output voltage to pull up to $V_{cc} - V_{BE}$ volts.

The maximum and minimum input and output voltages for logical 0 and 1 are, in the main, the same as those for the standard ttl gates. Differences are as follows: $V_{OH(min)} = 2.7$ V and $V_{OL(max)} = 0.5$ V.

A comparison between the various types of ttl gates is given by Table 2.1 which shows typical values for the parameters quoted.

CMOS Logic

The **complementary metal-oxide semiconductor** or **cmos** logic family offers the desirable features of very low power dissipation and good noise immunity. Its main disadvantage is its relatively long propagation delay produced by the time constants, arising from the very high input impedance of an enhancement-mode mosfet.

The circuit of a **cmos NAND gate** is given in Fig. 2.22. It can be seen that the n-channel mosfets are connected in series and the p-channel devices are in parallel. The operation of the circuit is as follows. If either, or both, of the inputs is at logical 0 ($\simeq 0 \text{ V}$), then the associated p-channel mosfets, T_1 and/or T_2, are turned ON, while the associated n-channel mosfets, T_3 and/or T_4, are turned OFF. The output terminal of the circuit is then at +5 V minus the saturation voltage of an ON mosfet. Conversely, if both inputs are at logical 1 ($\simeq 5 \text{ V}$), T_1 and/or T_2 are turned OFF, and T_3 and/or T_4 are turned ON. The output of the circuit is then at approximately 0 V or logical 0.

Fig. 2.22 CMOS NAND gate

Fig. 2.23 CMOS NOR gate

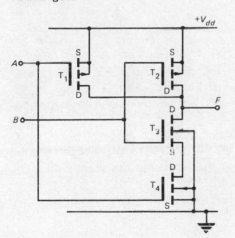

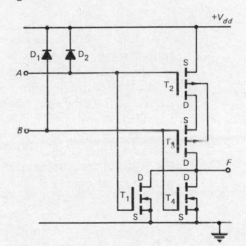

Very often *protective diodes* are connected between the input terminals and earth to reduce the possibility of the device being damaged by static charges produced by handling. The protective diodes are shown in the circuit of Fig. 2.23. Note that now the p-channel mosfets are connected in series and the n-channel mosfets are connected in parallel. If either of the input terminals is at logical 1, the associated p-channel mosfet (T_2 and/or T_3) is turned OFF and the associated n-channel mosfets (T_1 and/or T_4) are turned ON. The output voltage of the circuit is then low so that the NOR function is performed. Only if both inputs are at logical 0 will both the n-channel devices turn OFF and the p-channel mosfets turn ON so that the output can reach its logical 1 state.

Fig. 2.24 Transmission gate

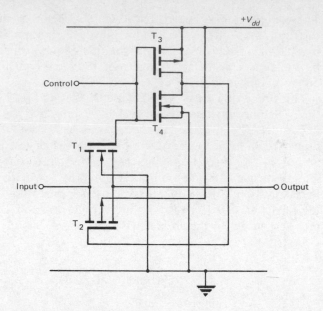

Another useful circuit available in the cmos family is known as the **transmission** or **analogue gate** (Fig. 2.24). The transistors T_1 and T_2 act as switches that can be turned ON (shut) or OFF (open) by a control signal applied to their gate terminals. One of the gate voltages should be the inverse of the other but this requirement is satisfied by means of the inverter provided by transistors T_3 and T_4.

CMOS devices are available in both the A and the B series, the B series operating from drain supply voltages of 3–20 V as opposed to 3–15 V for the A series. Also some of the electrical characteristics are slightly different. Some typical figures for cmos devices are

Low output voltage $V_{OL} = 0.05$ V max.
High output voltage $V_{OH} = V_{dd} - 0.05$ V min.
Noise margin 1 V min.
Propagation delay 30 ns
Sink current $I_{OL}(V_{dd} = 5 \text{ V}) = 1$ mA
Sink current $(V_{dd} = 15 \text{ V}) = 6.8$ mA
Source current $(V_{dd} = 5 \text{ V}) = 1$ mA
Source current $(V_{dd} = 15 \text{ V}) = 6.8$ mA

Unfortunately, the packing density of cmos devices is limited by the need for source-drain isolation and many lsi circuits use *either* p-channel mosfets *or* n-channel mosfets only.

PMOS Logic

A **pmos** integrated logic circuit uses *only* p-channel enhancement-mode mosfets and this means that the fabrication process is relatively simple. As a result, lsi devices can be manufactured with a high packing density and consequent low cost. Figs. 2.25*a* and *b* show the circuits of 2-input pmos

Fig. 2.25 (a) PMOS
NOR gate, (b) PMOS
NAND gate

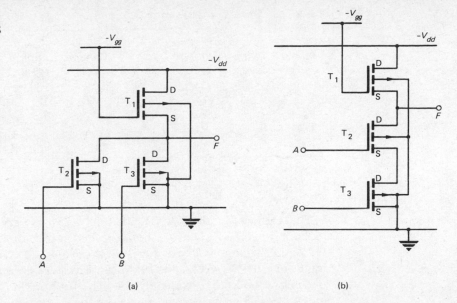

Fig. 2.25 (a) PMOS NOR gate, (b) PMOS NAND gate

NOR and NAND gates respectively. Since the drain supply voltage must be negative with respect to earth, it is usual to employ negative logic, i.e. logic 1 is represented by the most negative voltage. T_1 is biased, by the steady voltage $-V_{gg}$, to act as the *active load* resistor for the other transistors.

A p-channel enhancement-mode mosfet requires its gate to be held at a negative potential relative to its source for a drain current to flow. If the gate-source voltage is zero, the mosfet will be OFF.

Consider the **NOR gate** of Fig. 2.25a; if either or both inputs are low (at $-V_{dd}$ volts), or logical 1, then the associated transistor will conduct current and the output voltage will be at the saturation voltage of a mosfet ($\simeq -1\,V$), or logical 0.

When both input are high ($\simeq -1\,V$), or logical 0, both mosfets are OFF and the output voltage is $-V_{dd}$ volts or logical 1. The truth table for the circuit is given by Table 2.2. It is evident that the NOR function is performed.

Table 2.2   PMOS NOR gate

A		B		F	
Voltage level (V)	Logic level	Voltage level (V)	Logic level	Voltage level (V)	Logic level
$-V_{dd}$	1	$-V_{dd}$	1	-1	0
$-V_{dd}$	1	-1	0	-1	0
-1	0	$-V_{dd}$	1	-1	0
-1	0	-1	0	$-V_{dd}$	1

Table 2.3 PMOS NAND gate

A Voltage level (V)	Logic level	B Voltage level (V)	Logic level	F Voltage level (V)	Logic level
-1	0	-1	0	$-V_{dd}$	1
$-V_{dd}$	1	-1	0	$-V_{dd}$	1
-1	0	$-V_{dd}$	1	$-V_{dd}$	1
$-V_{dd}$	1	$-V_{dd}$	1	-1	0

The operation of the **NAND gate** (Fig. 2.25*b*) is very similar. When either or both inputs are high (0), the associated transistor does not conduct and the output voltage is low at $-V_{dd}$ volts or logical 1. Only when both of the inputs are at $-V_{dd}$ volts and both transistors are ON will the output voltage of the gate be at approximately 0 V. The truth table for the circuit is given by Table 2.3 from which it can be seen that the NAND function is performed.

Typically, a pmos device will have a power dissipation of 6 mW per gate and a propagation delay of 140 ms.

NMOS Logic

The circuitry of **nmos** logic gates is the same as shown in Fig. 2.25 for pmos except that n-channel devices are used and the polarities of the power supplies are reversed. NMOS devices are harder to manufacture than pmos and hence they tend to be more expensive, but they do possess the important advantages of being faster and of dissipating less power.

Both nmos and pmos techniques are used for lsi circuits such as memories and microprocessors but nmos has now replaced pmos for most new circuits because of its higher speed, higher density, and ttl-compatible voltage levels. The latest versions of nmos are known as hmos—the h standing for "high performance" since they provide an improvement in the speed-power product of about four times.

Emitter-coupled Logic

The main feature of **emitter-coupled logic,** or **ecl**, is its very fast speed of operation with which only Advanced Schottky ttl is a serious competitor. The ecl logic family generally finds application when the maximum possible speed of operation is the prime consideration. Very fast switching is obtained by ensuring that the transistors do not saturate when turned ON.

The basic gate in the ecl family is a combined **OR/NOR gate,** the circuit of which is given in Fig. 2.26. It should be noted that the circuit operates from a -5.2 V supply and logic 1 is represented by -0.75 V and logic 0 by -1.60 V. A reference voltage of -1.29 V is developed by R_7, D_1, D_2 and R_8 and is applied to the base terminal of transistor T_4.

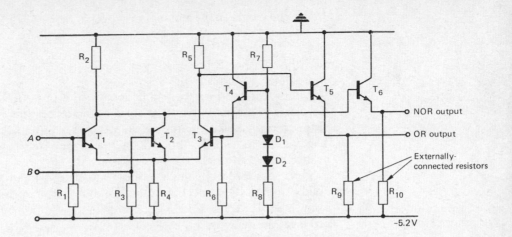

Fig. 2.26 ECL OR/NOR gate

If both inputs A and B are at logical 0 (-1.60 V), the base potential of T_3 will be less negative than the base potentials of both T_1 and T_2. Hence transistor T_3 conducts while T_1 and T_2 do not conduct. The common-emitter voltage is then equal to $-1.15 - 0.7 = -1.85$ V, and so an emitter current of $(-1.85+5.2)/R_4$ flows. Neglecting the base current of T_3, this is also the collector current of T_3 and it develops a voltage of

$$(-1.85+5.2)R_5/R_4$$

or about -1.0 V, across R_5, and this is sufficient to make T_5 conduct. T_5 is connected as an emitter follower and so its output voltage is

$$[(5.2-1.85)R_5/R_4] - V_{BE5} \text{ volts} \simeq -1.6 \text{ V or logical } 0$$

The collector potentials of T_1 and T_2 are at approximately 0 V and hence the output voltage of T_6 is

$$0 \text{ V} - V_{BE6} \simeq -0.75 \text{ V or logical } 1$$

If either input A or input B is at logical 1, i.e. -0.75 V, the base of the associated transistor is less negative than the base of T_3 and this transistor turns ON while T_3 turns OFF. The voltage dropped across R_5 is now zero and so the output voltage of T_5 switches to very nearly -0.7 V. The voltage developed across R_4 is

$$-0.75-0.7=-1.45 \text{ V} \quad \text{and} \quad I_E \simeq I_C = (-1.45+5.2)/R_4$$

and so the voltage across R_2 is

$$+3.75R_2/R_4 \quad \text{and} \quad V_{out(T6)}=V_{R2}-V_{BE6}$$

The noise margin of the circuit is easily determined:
a) When all the inputs are at logical 0 or -1.60 V, the voltage at the emitter of an input transistor is -1.75 V so that V_{BE} is 0.15 V. For a transistor to start conducting current, $V_{BE} \simeq 0.5$ V so that the 0 noise margin is 0.35 V.

b) When one of the inputs is at logical 1 or -0.75 V, then

$$V_{BE4} = -1.0 + 1.45 = 0.45 \text{ V}$$

and this is the 1 noise margin. Note that the two noise margins are approximately equal at 0.4 V.

Because of the high switching speeds, the fan-out is not determined by d.c. current effects but by the total capacitance that is present across the output terminals, i.e. input capacitance of a gate times the number of gates. The fan-out is usually about 50 at the lower frequencies but falls as the desired speed of operation is increased.

Some standard ecl gates are the 1660 dual 4-input OR/NOR, the 1662 quad 2-input NOR, the 1664 quad 2-input OR, and the 1672 triple 2-input exclusive-OR. These ics have a power dissipation of 60 mW per gate and a propagation delay of 1.1 ns.

Wired-OR logic is often used with ecl stages.

A variation of ecl is known as **current mode logic** or **cml**. CML uses voltage levels of 0 V for logical 1 and -0.5 V for logical 0 and does not employ output emitter follower stages. Because of this, cml circuits are able to operate from a collector supply voltage of only -3.3 V. As a result cml is as fast to operate as ecl but it has a lower speed-power product.

Comparison between the Integrated Random Logic Families

The main characteristics of the integrated logic families used for random logic are listed in Table 2.4. Typical figures are quoted.

Table 2.4 Integrated logic families

Family	Propagation delay (ns)	Power dissipation (mW)	Noise immunity (V)	Fan-in	Fan-out	Sink current (mA)	Source current (mA)
Standard ttl	9	10	0.4	8	10	16	0.8
Schottky ttl	3	20	0.3	8	10	8	0.4
Schottky low-power ttl	7	2	0.3	8	10	8	0.4
cmos	30	0.001	1.5	8	50	1.0	1.0
ecl	1.1	40	0.4	5	25		

Charge-coupled Devices

Fig. 2.27*a* shows a p-type substrate on the top of which has been grown a narrow layer of silicon dioxide. Positioned on top of this insulating layer is a metal plate.

If a positive potential, greater than the threshold voltage V_t, is applied to the plate (Fig. 2.27*b*), a depletion layer or region will be formed in the

Fig. 2.27 Formation of an inverse layer

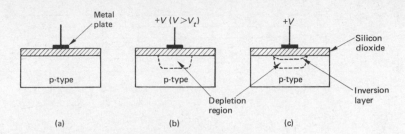

substrate. The dotted line represents the *surface potential* which is equal to the voltage applied to the plate *minus* the threshold voltage. Usually, $V_t \simeq 2\,\text{V}$. Any free electrons in the depletion region, produced either by thermal agitation, or injected in a way to be described later, will be attracted to the area immediately beneath the plate. The electrons are then said to be stored within a **potential well** and form an *inversion layer* (Fig. 2.27c).

The **charge-coupled device** works by moving the charge stored in one potential well to another potential well, and this means that it must contain a large number, perhaps several hundreds, of metal plates or *gates*. Consider Fig. 2.28a which shows a p-type substrate with four metal plates mounted onto its silicon dioxide insulating layer. Three of the plates have a positive

Fig. 2.28 Principle of a charge-coupled device

potential of +5 V applied to them but the second plate from the left has +10 V applied to it. All the available electrons, or charge, will be stored in the potential well beneath the second plate.

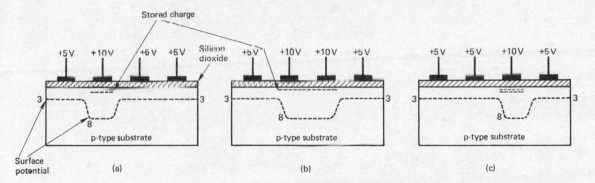

If the voltage applied to the third plate is increased to +10 V (Fig. 2.28b), the potential well beneath this plate will increase to the same depth as the well beneath the second plate. As a result the charge stored will be shared between the two potential wells as shown. Reducing the potential of the second plate to +5 V volts (Fig. 2.28c) means that the third plate is now at the most positive potential and so all the stored charge will be moved one place to the right into the potential well beneath the third plate. It should be clear that if the voltage on the fourth plate is now increased to +10 V, and then the potential of the third plate is reduced to +5 V, the stored charge can be moved one more place to the right.

A practical charge-coupled device may have hundreds of gates and it is not possible to connect an individual positive potential to each one of them

Fig. 2.29 3-phase ccd system

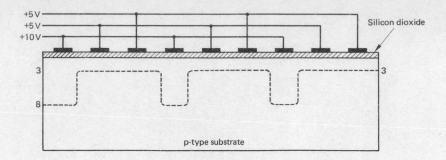

in turn. Instead, every third gate is commoned together (Fig. 2.29) to form a *three-phase system*. Clock waveforms of suitable phasing must be applied to each of the three lines in order to shift the charge along the device.

The method of injecting a packet of charge into the device varies; one method consists of applying a voltage pulse to the input gate that is of sufficiently high value to produce a momentary avalanche breakdown of the depletion region beneath the gate. For a binary 1 to be handled, the breakdown must occur at the instant that a potential well is present and is just about to move to the right. If no breakdown occurs as a well commences movement, then a binary 0 will be stored. Another method uses an input p-n junction; the p-type region is the substrate and the n-type region is a heavily-doped area into which electrons are injected by the application of a negative voltage (Fig. 2.30). Charge is placed into the potential well underneath the first transfer gate by applying a voltage greater than V_t to the input gate. The input diffusion region is normally held slightly reverse-biased and its function is to control the amount of charge placed into the well.

Fig. 2.30 Complete ccd system

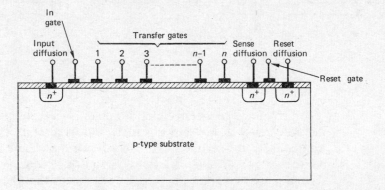

The arrival of charge packets at the output of the device is detected by connecting the sense diffusion electrode to a positive potential via a load resistor. The sense diffusion electrode acts as a reverse-biased diode. The received charge packets cause small currents to flow in the load resistor and develop a voltage across it. After the arrival of each packet the potential of the sense diffusion electrode must be reset. The reset electrode is held at a more positive potential than the sense diffusion electrode. To reset the sense diffusion electrode the reset gate has a voltage pulse applied to it after each charge packet has been detected and this moves the received charge to the reset electrode.

Silicon Gates

When a large number of gates are to be provided, and a typical ccd may well have several hundreds, difficulties arise with the fabrication of aluminium gates with *very* narrow gaps between them. The use of metallization has therefore been replaced, for most devices, by **silicon gates**.

Fig. 2.31 Silicon gate

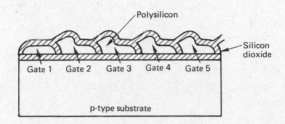

The principle of a silicon gate is shown by Fig. 2.31. A layer of polysilicon (an insulator) is deposited on top of the silicon dioxide layer and is then etched to form one third of the required number of gates, i.e. every third one. A layer of silicon oxide is formed and is then etched away as appropriate to form the remaining gates, i.e. 3, 6, 9, etc. and then, finally, a layer of silicon dioxide is grown overall. The separation between adjacent gates is now only the thickness of a silicon dioxide layer and is much less than could otherwise be obtained.

Many commercially available ccds employ one or both of two variations of the basic scheme outlined. *Two-phase* devices have the advantage of needing one less clock source but are more complicated to manufacture, and *buried-channel* devices move the charge packets through the body of the substrate rather than over its surface.

Integrated Injection Logic

Integrated injection logic or **I²L** is a method of using integrated bipolar transistors without the need for each element to be separately isolated. The packing density that can be achieved is greater than even mosfet technology can achieve, and hence I²L logic is eminently suitable for use in lsi devices. One of the main advantages of I²L is that it is easily combined in the same chip, with any of the other logic families.

I²L avoids the complexity of conventional bipolar transistor logic by the use of **inverted transistors**. This term simply means that the emitter and collector regions of a transistor are interchanged. A consequence of this is that I²L transistors can have multiple collectors and so wired-OR logic is convenient.

Fig. 2.32 I^2L n-p-n transistor

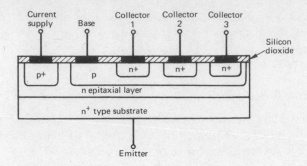

Fig. 2.32 shows the fabrication of an I^2L n-p-n transistor with three collectors. The emitter of the transistor is the n epitaxial layer; this layer extends throughout the chip and provides the emitter for *every* n-p-n transistor formed within the chip.

p-n-p transistors are used as current sources and these are of the *lateral* type [EIV]. The terminal marked as "current supply" is the emitter, the terminal marked "emitter" is the base, and the terminal labelled as "base" becomes the collector.

Fig. 2.33 I^2L OR/NOR gate

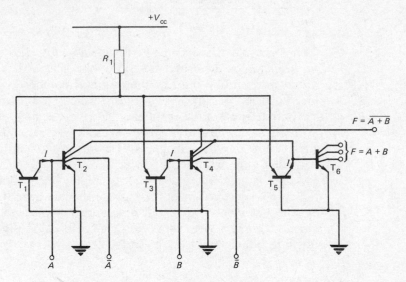

Fig. 2.33 shows a 2-input **OR/NOR gate** which also gives inversion of each input. The common base transistors T_1, T_3, and T_5 are operated as constant current sources that deliver a constant current to the base of each n-p-n transistor. The magnitude of the constant current is determined by the value chosen for an externally connected resistor R_1. Only the one resistor is used to bias *all* the current sources within the ic and there could be several thousands of them.

Suppose both inputs are at logical 0 or approximately 0.05 V. Transistors T_2 and T_4 are then OFF and $\bar{A}$ and $\bar{B}$ outputs are high at the logical 1 voltage level of 0.75 V. The $F = \overline{A + B}$ output terminal is also high. T_6 is turned full ON and its output is at logical 0.

When one of the inputs, say A, is at logical 1 and the other input, B, is at logical 0, transistor T_1 will be saturated and T_4 will be OFF. Since the collectors of T_1 and T_4 are connected together in wired-OR, the output $\overline{A+B}$ must be at logical 0. Also the output of the inverting stage T_6 must be at logical 1.

Only when both the inputs are at logical 1 will both T_1 and T_4 be saturated so that the output of the circuit will be low. Now T_6 will be OFF and its output will be high.

Fig. 2.34 I^2L
AND/NAND gate

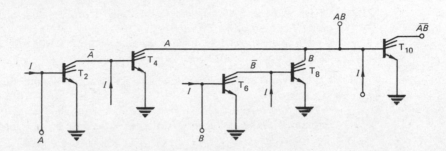

To obtain the **AND** and **NAND** functions using I^2L logic, the transistors T_2, T_4 and T_6 must be interconnected differently. Fig. 2.34 shows the connections needed to produce AND and NAND gates; the current sources have been omitted to simplify the drawing.

The basic gate, a p-n-p transistor current source and an n-p-n inverter, can be interconnected in a number of different ways to generate almost any required logical function.

I^2L offers a number of advantages over other lsi techniques. These advantages are: a) higher packing density, b) compatibility with ttl circuitry, c) low power dissipation/operating speed product, d) faster operation than mosfet technology.

I^2L is used in the manufacture of some microprocessors.

Exercises 2

2.1 a) Describe, with the aid of a circuit diagram, the operation of a ttl 3-input NAND gate.

b) Write down the truth table for the circuit given and state how the logical function of the circuit changes if negative logic is used.

c) How does ttl compare with cmos in respect of (i) noise immunity, (ii) switching speed, (iii) power consumption, (iv) fan-out?

2.2 With the aid of a circuit diagram describe the operation of a 4-input ttl NOR gate using positive logic. Explain what limits the switching speed.

2.3 State the properties of an *ideal* switch and then compare with the properties of a bipolar transistor switch. Sketch typical output characteristics for a bipolar transistor operated as a switch. Draw a suitable load line and show clearly the ON and OFF regions. Label the voltage $V_{CE(sat)}$. Explain how the power dissipated by the transistor varies during the switching cycle OFF-ON-OFF.

2.4 Describe, with the aid of a circuit diagram, the operation of a 3-input cmos NAND gate.

Draw up a table to show how the speed, power dissipation and other parameters of Schottky low-power ttl compare with cmos logic.

2.5 With the aid of a circuit diagram and a truth table explain the operation of an ecl OR/NOR gate.

In the circuit of Fig. 2.26, $R_2 = 290\Omega$, $R_4 = 1180\Omega$, $R_5 = 300\Omega$. If the reference voltage at the base of T_4 is -1.29 V, calculate the output voltage of the circuit a) when it is high, b) when it is low. Take logic 1 as -0.75 V and logic 0 as -1.60 V.

2.6 a) Draw typical static characteristics for a semiconductor diode. Explain how a diode can be used as a switch.

b) Explain how the diode parameters (i) forward recovery time, (ii) reverse recovery time are affected by high-frequency effects.

2.7 Explain with the aid of the static characteristics of a diode how a diode can be used as a switch.

The voltage pulse shown in Fig. 2.35a is applied to the circuit of Fig. 2.35b. Explain, with the aid of suitable diagrams, the voltage that appears across the diode.

Fig. 2.35

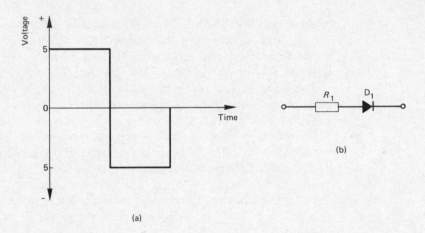

(a)

(b)

2.8 a) Describe, with the aid of characteristics, how (i) a bipolar transistor, (ii) a mosfet can be used as a switch.

b) Compare the relative advantages and disadvantages of these devices as switches.

2.9 Briefly explain the operation of a bipolar transistor as a switch. Explain the meaning of the term *hole storage time* and how it arises. Explain why the hole storage time may be a disadvantage in a switching transistor and say how it can be reduced.

2.10 Explain the purpose of the totem pole output stage used in a ttl gate. The transistor used in the circuit of Fig. 2.36 has the following parameters: $V_{CE(sat)} = 0.2$ V and $V_{BE(sat)} = 0.6$ V. Calculate a) the current flowing in the transistor T_3 when the output is low if the diode D_1 were not present, b) the minimum current gain for T_3 for the gate to be able to change state.

Fig. 2.36

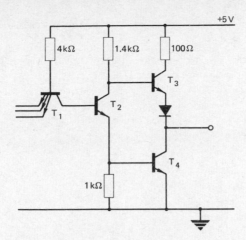

2.11 Describe briefly, with the aid of sketches, the operation of *a*) a cmos logic gate, *b*) an I²L logic gate.

Compare these two forms of logic for their suitability for use in lsi.

2.12 Draw the circuit of a dtl NAND gate and quote typical values for each component shown. If $h_{FE} = 50$, determine the maximum fan-out of the circuit.

2.13 A switching transistor has the following data:

$V_{CB(max)} = 20$ V, $V_{CE(max)} = 20$ V, $I_{C(max)} = 450$ mA, $P_{Diss(max)} = 520$ mW
$V_{CE(sat)} = 250$ mV, $V_{BE(sat)} = 800$ mV
Rise time $= 22$ ns   Storage time $= 100$ ns
Delay time $= 6$ ns   Fall time $= 28$ ns

Draw some typical characteristics and hence show how the given ratings limit the choice of supply voltage and collector load resistance.

Explain the meaning of each of the four switching times mentioned.

Short Exercises

2.14 The rising output voltage of a low-power Schottky ttl gate is given by

$$V(t) = V_{OL} + 3.7(1 - e^{-t/8 \times 10^{-9}})$$

State the meaning of the symbol V_{OL} and quote a typical value. Use this value to determine the time taken for the output voltage to rise to 90% of $V_{cc} = 5$ V.

2.15 What is meant by saturation in a bipolar transistor circuit? A transistor has a collector resistance of 2.2 kΩ and a supply voltage of 12 V. Calculate the saturated collector current if $V_{CE(sat)} = 0.2$ V. If $V_{BE(sat)} = 0.75$ V determine the forward bias voltage of the collector/base junction.

2.16 Define the terms fan-in, fan-out, operating temperature range, and noise margin with reference to logic gates.

2.17 Draw a totem pole output stage and explain why two such stages should never have their outputs connected in parallel.

2.18 Describe, with the aid of the circuit diagram, the operation of a 3-input cmos NAND gate.

2.19 With the aid of a circuit diagram explain the operation of a 2-input low-power Schottky NAND gate.

2.20 If the maximum collector current rating of the transistor T_2 in Fig. 2.20 is 30 mA determine the maximum fan-out of the circuit.

3 Combinational Logic Circuits

Introduction

Many digital circuits can be constructed using a number of suitably interconnected logic elements such as gates and bistable multivibrators (see Chapter 4). If a circuit does not involve a need for time delay or storage, only gates of one kind are necessary, Many of the more common digital circuits, such as binary adders for example, are packaged as integrated circuit devices in the various logic families.

In this chapter the various ways in which gates can be interconnected to produce more complex digital functions will be described and, where appropriate, details of corresponding msi or lsi circuits will be given.

Binary Adders

Very often in digital circuitry the need arises for two binary numbers to be added together to produce a **sum** and a **carry**. If there is no carry from a previous stage a half-adder will suffice, but if a previous carry must be taken into account then a full-adder is required.

The design process for a digital circuit starts with the writing down of the truth table of the required logical operation. Each 1 that appears in the output column of the truth table must be represented by a term in the Boolean equation that describes the operation of the circuit. The Boolean equation must contain each literal that is at 1 and the complement of each literal that is at 0.

The Half-adder

The truth table of a **half-adder** is given by Table 3.1. Two different Boolean equations can be obtained from Table 3.1 that express the sum and the carry. These equations are

$$S = A\bar{B} + \bar{A}B \qquad C = AB \qquad\qquad (3.1)$$
$$S = (A + B)\overline{AB} \qquad C = AB \qquad\qquad (3.2)$$

Table 3.1 Half-adder truth table

A	0	1	0	1
B	0	0	1	1
Sum	0	1	1	0
Carry	0	0	0	1

These equations can be directly implemented using a combination of AND and OR gates and inverters, but for the reasons given in Chapter 1 it is more likely that either NAND or NOR gates would be exclusively employed. Since equation (3.1) is in the sum-of-products form it is best suited to a NAND gate implementation; while equation (3.2) is easier to implement using NOR gates.

Fig. 3.1 Half-adder circuits (a) and (b) using NAND gates only, (c) using NOR gates only

Fig. 3.2 The circuit of Fig. 3.1a implemented using the LS 7400 quad 2-input NAND gate

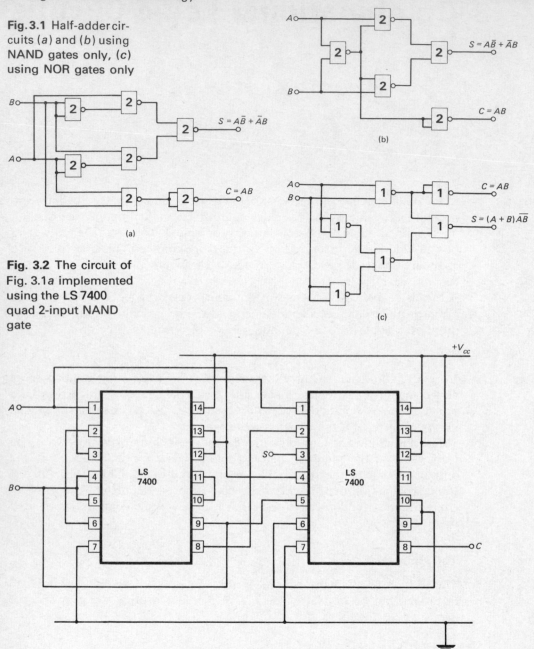

Using the rules enunciated on page 18, Figs. 3.1a and c can be drawn directly. A simpler alternative to Fig. 3.1a is given in b.

As an example of how these circuits could be fabricated using ttl or cmos devices, Fig. 3.2 shows Fig. 3.1a implemented using the low-power Schottky ttl integrated circuit LS 7400 quad 2-input NAND gate.

Table 3.2 Full-adder truth table

A	0	1	0	0	1	1	0	1
B	0	0	1	0	1	0	1	1
C_{in}	0	0	0	1	0	1	1	1
Sum	0	1	1	1	0	0	0	1
C_{out}	0	0	0	0	1	1	1	1

Full-adder

The truth table of a **full-adder** is given by Table 3.2. From it, Boolean expressions for the sum S and the output carry C_{out} can be obtained:

$$S = A\bar{B}\bar{C}_{in} + \bar{A}B\bar{C}_{in} + \bar{A}\bar{B}C_{in} + ABC_{in}$$

$$= (A\bar{B} + \bar{A}B)\bar{C}_{in} + (\bar{A}\bar{B} + AB)C_{in} \tag{3.3}$$

$$= (\text{exclusive-OR})\,\bar{C}_{in} + (\text{exclusive-NOR})C_{in} \tag{3.4}$$

Also,

$$C_{out} = AB\bar{C}_{in} + A\bar{B}C_{in} + \bar{A}BC_{in} + ABC_{in}$$

$$= AB(\bar{C}_{in} + C_{in}) + C_{in}(A\bar{B} + \bar{A}B)$$

$$= AB + C_{in}(\text{exclusive-OR}) \tag{3.5}$$

The expressions obtained for the sum and the carry-out can be implemented using NAND or NOR gates only, or use can be made of exclusive-OR gates. The NAND gate implementation of the full-adder is shown in Fig. 3.3.

Fig. 3.3 Full-adder using NAND gates only

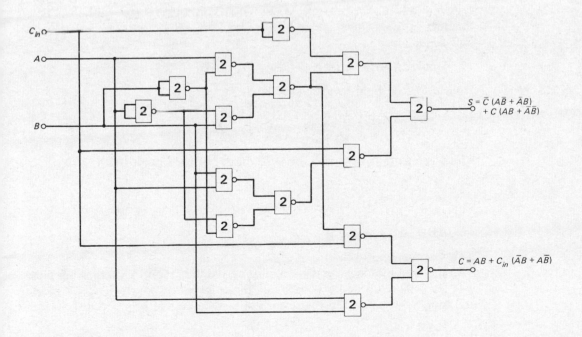

Fig. 3.4 Full-adder using exclusive-OR and NAND gates

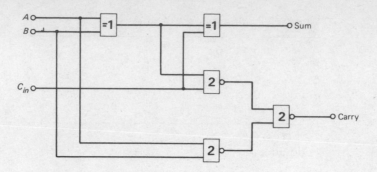

Equation (3.4) can be implemented, somewhat more simply, by using exclusive-OR gates, e.g. the 7486 quad 2-input, connected as shown by Fig. 3.4. It is not obvious that this circuit achieves the desired result but the output of the right-hand exclusive-OR gate is

$$S = (A\bar{B} + \bar{A}B)\bar{C}_{in} + C_{in}(AB + \bar{A}\bar{B})$$

and

$$C = \overline{AB\ C_{in}(A\bar{B} + \bar{A}B)}$$
$$= AB + C_{in}(A\bar{B} + \bar{A}B)$$

Fig. 3.5 Connection of two half-adders to produce one full-adder

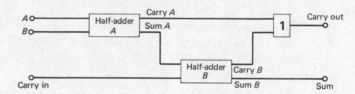

A full-adder can also be constructed by connecting two half-adders together in the manner shown by Fig. 3.5. The two outputs of the left-hand half-adder are

$$\text{Sum } A = A\bar{B} + \bar{A}B \quad \text{and} \quad \text{Carry } A = AB$$

Hence, for the right-hand half-adder,

$$\text{Sum } B = (A\bar{B} + \bar{A}B)\bar{C}_{in} + \overline{(A\bar{B} + \bar{A}B)}C_{in}$$
$$= (A\bar{B} + \bar{A}B)\bar{C}_{in} + (AB + \bar{A}\bar{B})C_{in}$$

which is equation (3.4) and the sum output of the circuit. Also,

$$\text{Carry } B = (A\bar{B} + \bar{A}B)C_{in}$$

This means that the carry-out of the circuit is

$$AB + (A\bar{B} + \bar{A}B)C_{in}$$

which is equation (3.5).

Complete full-adder circuits are also available in the ttl and cmos families. The ttl 7482 2-bit adder consists of three inverters, fourteen AND gates, and four NOR gates, showing clearly the advantage of using an msi device.

Fig. 3.6 4-bit full-adder circuits (*a*) using two 7482 2-bit full-adders, (*b*) using one 7483 4-bit full-adder

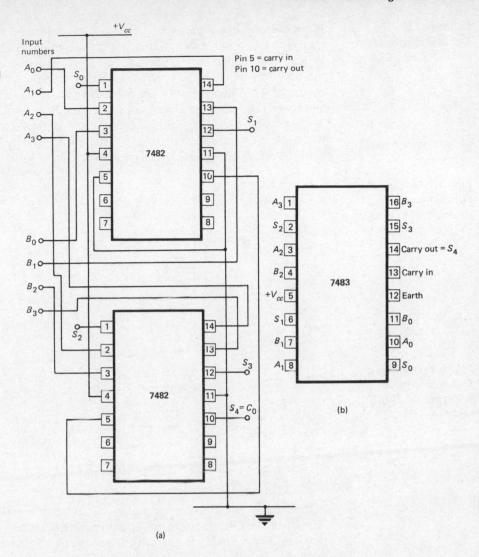

If two 4-bit numbers are to be added, two 7482s can be interconnected as shown by Fig. 3.6*a*, but it would be easier and cheaper to use a 4-bit adder like the 7483 (Fig. 3.6*b*). Note that the carry-in pin is connected to earth when there is no input carry. In both circuits A_0 and B_0 are the least significant bits.

This method of binary addition is known as **parallel addition** since each bit is handled by separate circuitry.

The parallel adder requires more circuitry than the alternative, known as the **serial adder,** but its speed of operation is very much greater. The only time delay that occurs before the result of the addition is obtained is determined by the need for the carry to propagate through the adder. Increased speed of operation at increased cost is possible if logic circuitry is added to produce "carry look-ahead"

Magnitude Comparators

A **magnitude comparator** is a circuit that determines whether one binary number A is greater than, equal to, or less than another binary number B. When only single-bit numbers are involved, the exclusive-OR gate can be used to decide whether or not two binary numbers A and B are equal to one another. The output of an exclusive-OR gate will be at logical 0 only when both of its inputs are equal, i.e. if $A = B$. If two n-bit numbers A and B are to have their magnitudes compared, n exclusive-OR gates can be used whose outputs fan into a single NOR gate to give a logical 1 output only when $A = B$ (see Fig. 3.7 which shows the circuit for the case when $n = 4$).

Fig. 3.7 Magnitude comparator using exclusive-OR gates (indicates $A = B$ or $A \neq B$ only)

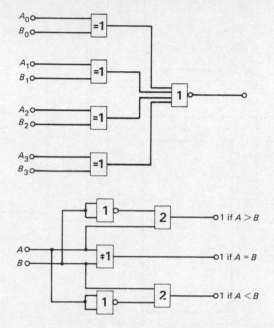

Fig. 3.8 1-bit magnitude comparator (indicates $A = B$, $A > B$, or $A < B$)

If, when A and B are not equal, some indication is required as to whether $A > B$ or $A < B$, then extra circuitry is necessary. Fig. 3.8 shows one possible arrangement for a 1-bit comparator. The same principle *can* be extended to allow numbers containing any number of bits to have their magnitudes compared but even for a 4-bit comparator the circuitry rapidly becomes very complicated. 4-bit magnitude comparators are available in both the ttl (7485) and the cmos (4063) families. These two devices have identical pin connections and these are shown in Fig. 3.9a. If the two numbers to be compared contain more than four bits, two or more ics will be needed. Fig. 3.9b shows how two 7485 4-bit magnitude comparators are connected together to enable two 8-bit numbers to be compared. It can be seen that the $A > B$, $A = B$, and $A < B$ outputs of the least significant comparator (i.e. the one whose inputs are A_0, A_1, A_2, A_3 and B_0, B_1, B_2, B_3) are connected to the similarly labelled inputs of the next significant stage. The least significant stage must have its $A = B$ input terminals connected to $+V_{cc}$ volts and its $A > B$ and $A < B$ terminals connected to earth.

Fig. 3.9 (*a*) Pin connections of two 4-bit magnitude comparators

   (*b*) an 8-bit magnitude comparator

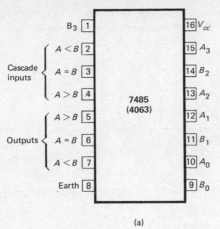

(a)

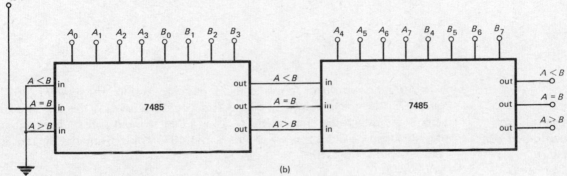

(b)

Code Converters

Many instances arise for a decimal number to be encoded into the corresponding binary or binary coded decimal (or some other code) number. Similarly, it is often necessary to decode a number from pure binary or bcd into decimal.

Decimal-to-Binary Coded Decimal Converter

A **decimal-to-bcd converter** will have nine input lines representing respectively the decimal integers $0, 1, 2, 3, 4, 5, 6, 7, 8, 9$, and four output lines representing $2^0, 2^1, 2^2$ and 2^3 respectively. The presence of any one of the nine decimal numbers is indicated by a high level (logic 1 voltage) on the appropriate input line and a low level (logic 0 voltage) on all of the remaining 1 input lines. Decimal zero is represented by *all* the input lines being at the low level.

The converter is required to generate the logic 1 state on the appropriate output lines to produce the binary equivalent of the decimal input number. The truth table of a decimal-to-bcd converter is given by Table 3.3.

Table 3.3 Decimal-to-bcd truth table

Decimal		0	1	2	3	4	5	6	7	8	9
BCD	2^0	0	1	0	1	0	1	0	1	0	1
out-	2^1	0	0	1	1	0	0	1	1	0	0
puts	2^2	0	0	0	0	1	1	1	1	0	0
	2^3	0	0	0	0	0	0	0	0	1	1

From Table 3.3 the Boolean equation describing the operation of the circuit is

$$2^0 = 1 + 3 + 5 + 7 + 9$$
$$2^1 = 2 + 3 + 6 + 7$$
$$2^2 = 4 + 5 + 6 + 7$$
$$2^3 = 8 + 9$$

Fig. 3.10 Two decimal-to-bcd converters

If OR gates with two, four and five inputs are available, e.g. 4071, 4072 and 4078, the circuit given in Fig. 3.10a could be used at the expense of using three integrated circuits (the 4072 is a dual 4-input gate). The number of ics needed can be reduced to two if the 4075 triple 3-input OR gate is used instead (Fig. 3.10b).

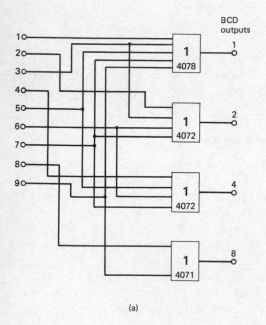

(a)

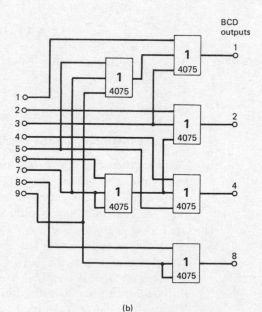

(b)

Fig. 3.11 BCD-to-decimal converter

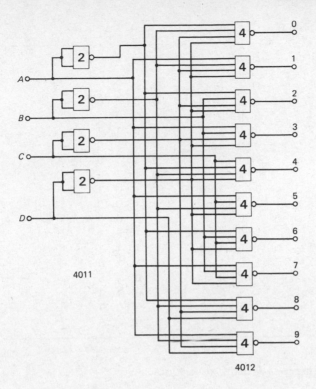

4011

4012

The design of a **bcd-to-decimal converter** can also start from the truth table given by Table 3.3. Let $2^0 = A$, $2^1 = B$, $2^2 = C$ and $2^3 = D$. Then

$$0 = \bar{A}\bar{B}\bar{C}\bar{D} \qquad 1 = A\bar{B}\bar{C}\bar{D} \qquad 2 = \bar{A}B\bar{C}\bar{D} \qquad 3 = AB\bar{C}\bar{D} \qquad 4 = \bar{A}\bar{B}C\bar{D}$$

$$5 = A\bar{B}C\bar{D} \qquad 6 = \bar{A}BC\bar{D} \qquad 7 = ABC\bar{D} \qquad 8 = \bar{A}\bar{B}\bar{C}D \qquad 9 = A\bar{B}\bar{C}D$$

These equations can be implemented directly using four inverters and ten AND gates but an alternative circuit, shown by Fig. 3.11, uses NAND gates only. In this circuit, of course, the *presence* of a given decimal number is denoted by the binary 0 state. The circuit shown could be implemented using five dual 4-input NAND gates such as the 4012 and one quad 2-input NAND gate, e.g. the 4011. Alternatively, a hex inverter (4069) could replace the 4011. If the "don't care" numbers 10 through to 15 are taken into account the circuit can be implemented using only one dual 4-input NAND gate, two triple 3-input NAND gates and one quad 2-input NAND gate (Exercise 3.26).

Converters can be designed in similar manner for the conversion of a number from one code into any other code, e.g. pure binary into excess-3. When a bcd-to-decimal decoder is required it is both cheaper and more convenient to use an integrated circuit decoder, a number of which exist in both the ttl and the cmos families. Some examples are mentioned later.

Table 3.4

Deci-mal no.	Binary				Gray			
	D	C	B	A	G_4	G_3	G_2	G_1
0	0	0	0	0	0	0	0	0
1	0	0	0	1	0	0	0	1
2	0	0	1	0	0	0	1	1
3	0	0	1	1	0	0	1	0
4	0	1	0	0	0	1	1	0
5	0	1	0	1	0	1	1	1
6	0	1	1	0	0	1	0	1
7	0	1	1	1	0	1	0	0
8	1	0	0	0	1	1	0	0
9	1	0	0	1	1	1	0	1
10	1	0	1	0	1	1	1	1
11	1	0	1	1	1	1	1	0
12	1	1	0	0	1	0	1	0
13	1	1	0	1	1	0	1	1
14	1	1	1	0	1	0	0	1
15	1	1	1	1	1	0	0	0

Binary-to-Gray Converters

Table 3.4 shows the binary and the Gray code equivalents of the decimal numbers 0 through to 15. From the truth table,

$$G_1 = A\bar{B}\bar{C}\bar{D} + \bar{A}B\bar{C}\bar{D} + A\bar{B}C\bar{D} + \bar{A}BC\bar{D} + A\bar{B}\bar{C}D + \bar{A}B\bar{C}D + A\bar{B}CD + \bar{A}BCD$$

$$G_2 = \bar{A}B\bar{C}\bar{D} + AB\bar{C}\bar{D} + \bar{A}\bar{B}C\bar{D} + A\bar{B}C\bar{D} + \bar{A}B\bar{C}D + AB\bar{C}D + \bar{A}\bar{B}CD + A\bar{B}CD$$

$$G_3 = \bar{A}\bar{B}C\bar{D} + A\bar{B}C\bar{D} + \bar{A}BC\bar{D} + ABC\bar{D} + \bar{A}\bar{B}\bar{C}D + A\bar{B}\bar{C}D + \bar{A}B\bar{C}D + AB\bar{C}D$$

$$G_4 = \bar{A}\bar{B}\bar{C}D + A\bar{B}\bar{C}D + \bar{A}B\bar{C}D + AB\bar{C}D + \bar{A}\bar{B}CD + A\bar{B}CD + \bar{A}BCD + ABCD$$

Mapping on a Karnaugh map and simplifying gives

$$G_1 = \bar{A}B + A\bar{B} \qquad G_2 = \bar{B}C + B\bar{C} \qquad G_3 = \bar{C}D + C\bar{D} \qquad G_4 = D$$

There are several ways in which the terms for G_1, G_2, G_3 and G_4 can be implemented.

1 Using three 2-wide 2-input AOI gates and four inverters as shown by Fig. 3.12a. This implementation could be achieved with three ics; for example two dual 2-wide 2-input AOI gates such as the 7450 and one hex inverter such as the 7404. It is left as an exercise to confirm the operation of this circuit.

2 Using three exclusive-OR circuits and a hex inverter.

3 Using NAND or NOR gates or, of course, a combination of AND and OR gates and inverters.

A **Gray-to-binary converter** can be similarly designed by reading the truth table of Table 3.4 the other way around. Thus

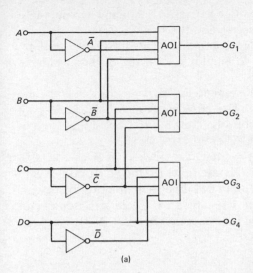

(a)

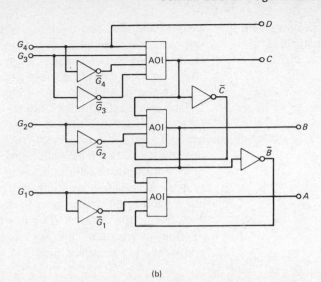

(b)

Fig. 3.12 (*a*) Binary-to-Gray code converter, (*b*) Gray-to-binary code converter

$$A = G_1\bar{G}_2\bar{G}_3\bar{G}_4 + \bar{G}_1 G_2\bar{G}_3\bar{G}_4 + G_1 G_2 G_3\bar{G}_4 + \bar{G}_1\bar{G}_2 G_3\bar{G}_4$$
$$+ G_1\bar{G}_2 G_3 G_4 + \bar{G}_1 G_2 G_3 G_4 + G_1 G_2\bar{G}_3 G_4 + \bar{G}_1\bar{G}_2\bar{G}_3 G_4 \tag{3.6}$$

This equation cannot be simplified by mapping.

$$B = G_1 G_2\bar{G}_3\bar{G}_4 + \bar{G}_1 G_2\bar{G}_3\bar{G}_4 + G_1\bar{G}_2 G_3\bar{G}_4 + \bar{G}_1\bar{G}_2 G_3\bar{G}_4$$
$$+ G_1 G_2 G_3 G_4 + \bar{G}_1 G_2 G_3 G_4 + G_1\bar{G}_2\bar{G}_3 G_4 + \bar{G}_1\bar{G}_2\bar{G}_3 G_4 \tag{3.7}$$

$$C = \bar{G}_1 G_2 G_3\bar{G}_4 + G_1 G_2 G_3\bar{G}_4 + G_1\bar{G}_2 G_3\bar{G}_4 + \bar{G}_1\bar{G}_2 G_3\bar{G}_4$$
$$+ \bar{G}_1 G_2\bar{G}_3 G_4 + G_1 G_2\bar{G}_3 G_4 + G_1\bar{G}_2\bar{G}_3 G_4 + \bar{G}_1 G_2\bar{G}_3 G_4 \tag{3.8}$$

Mapping and simplifying,

$$C = G_3\bar{G}_4 + \bar{G}_3 G_4 \tag{3.9}$$

$$B = G_2 G_3 G_4 + \bar{G}_2 G_3\bar{G}_4 + G_2\bar{G}_3\bar{G}_4 + \bar{G}_2\bar{G}_3 G_4 \tag{3.10}$$

$$D = G_4$$

The equations for *A*, *B*, and *C* can be implemented directly but considerable circuit simplification results if it is noticed that

$$B = \bar{G}_2 C + G_2\bar{C} \tag{3.11}$$

and $\quad A = G_1\bar{B} + \bar{G}_1 B \tag{3.12}$

(See Exercise 3.16.)

The AOI implementation of the Gray-to-binary converter is shown in Fig. 3.12*b*.

The ttl logic family includes some converters:
7442 bcd-to-decimal converter
7443 excess 3-to-decimal converter
74145 bcd-to-decimal converter
74184 bcd-to-pure binary converter
74185 binary-to-bcd converter.

Multiplexers and Demultiplexers

Two other combinational logic circuits often used in digital circuitry are known as the *multiplexer* or *data selector* and the *demultiplexer* or *decoder*. Both circuits can be constructed using random logic but since both are fairly complicated it would be more usual to employ one of the devices available in both the ttl and the cmos families.

The use of the multiplexer to implement logical functions has already been discussed in Chapter 1 (p. 23). The circuit can also be used to gain access to data that originates from more than one source, one application of this arising in the *read only memory* (Chapter 7).

The **multiplexer** selects one-out-of-n input lines, where n is usually 4, 8 or 16, and consists essentially of an AOI gate in which the NOR output gate has a number of inputs equal to the number of input data lines to the device. Any one of the input AND gates can be enabled by a data selection circuit so that input data can be selected and applied to the NOR gate by addressing the required input.

Fig. 3.13 Data selector

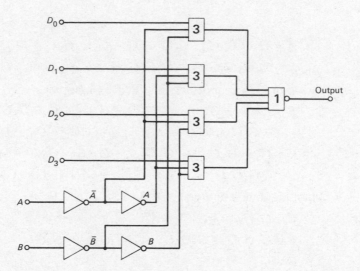

As a simple example, consider a 4-to-1 data selector. The data selection circuit must have two inputs A and B since $2^2 = 4$ and a suitable circuit is shown in Fig. 3.13. Each AND gate has three inputs, one of which is connected to a data input line, and the other two are connected to the selection circuit which merely consists of four inverters. Suppose the input address is $A = 0$ and $B = 1$. Then $\bar{A} = 1$ and $\bar{B} = 0$ and only the third AND gate from the top is enabled. The output of the circuit is then $\bar{D}_2$.

TTL data selectors include the 74151 8-to-1 line, the 74157/8 quad 2-to-1 line, the 74153 dual 4-to-1 line, and the 74150 16-to-1 line.

A **demultiplexer** or decoder performs the inverse function to that performed by a multiplexer. A single input line can be connected to any one of n output lines, the required output being selected by a binary address. The number of output lines is equal to 4, 8 or 16, with 2, 3 or 4 input address bits needed to allow any line to be selected.

Fig. 3.14 Decoder/multi-plexer

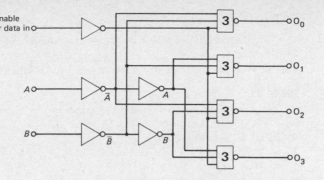

Consider a 1-to-4 line demultiplexer; two select lines A and B are needed and an *enable* or *data-in* input terminal. Four NAND gates are required, one for each output line, and any one of them can be selected by the address bits A and B but is only enabled when the enable input terminal is low. The circuit is shown by Fig. 3.14. When the enable terminal is low, the selected NAND gate will have a low output; all the other outputs will be high. If the enable terminal is kept low, the circuit will act as a 1-to-4 line *decoder* but, if this terminal is used as a data input terminal, the circuit will operate as a 1-to-4 line demultiplexer. The selected output line will then be low when the data input is low and high when the data input is high. The other outputs will remain high.

A number of decoders/demultiplexers are included in the ttl family: the 74138 3-to-8 line, the 74139 dual 2-to-4 line, and the 74154 4-to-16 line.

Diode Matrices

The conversion of numbers from one code to another can also be carried out using a **diode matrix.** Fig. 3.15 shows one example, a decimal-to-bcd converter. A bcd output line, A, B, C or D will be at the logical 1 voltage level whenever a decimal input line is at logical 1 *and* a diode is connected between the output line and that decimal input line.

For example,

$$A = 1+3+5+7+9$$
$$B = 2+3+6+7$$
$$C = 4+5+6+7$$
$$D = 8+9$$

When such a matrix is fabricated within an integrated circuit, each diode will probably be the base/emitter diode junction of an n-p-n transistor with the collector taken directly to $+V_{cc}$ volts.

The circuit of a diode matrix octal decoder is shown by Fig. 3.16. Its operation is left as an exercise (Exercise 3.13). A diode matrix is really a form of read-only memory and will be discussed further in Chapter 7.

Exercises 3

3.1   *a*) Using truth tables and logic diagrams draw the circuit and explain the operation of a full-adder.

   b) Draw a logic circuit to implement the truth table of Table 3.5.

Table 3.5

Input				Output			
A	B	C	D	W	X	Y	Z
0	0	0	0	1	1	0	0
0	0	0	1	1	0	1	1
0	0	1	0	1	0	1	0
0	0	1	1	1	0	0	1
0	1	0	0	1	0	0	0

3.2   A digital circuit has four binary input lines A, B, C, and D and a single output F. The truth table of the circuit is given by Table 3.6. Obtain the Boolean equation describing the logical operation of the circuit. Minimize the equation and draw the logic diagram of the circuit using either NAND or NOR gates only.

Table 3.6

A	0	0	0	0	0	0	0	0	1	1	1	1	1	1	1	1
B	0	0	0	0	1	1	1	1	0	0	0	0	1	1	1	1
C	0	0	1	1	0	0	1	1	0	0	1	1	0	0	1	1
D	0	1	0	1	0	1	0	1	0	1	0	1	0	1	0	1

3.3   Design a logic circuit to compare two 3-bit binary numbers A and B. The circuit should give outputs to indicate each of the conditions $A > B$, $A = B$ and $A < B$.

3.4   Derive expressions for the sum and the carry outputs of a full-adder. Show how these equations might be implemented. Discuss the advantages of using an msi package.

3.5   Show how the expression

$$f(A, B, C, D) = (4, 6, 12, 13, 14)$$

may be represented on a Karnaugh map and explain the procedure for obtaining the minimal sum of prime implicants.

3.6   Design, with the aid of a truth table and a Karnaugh map, a majority decision circuit that will give an output at logical 1 whenever two out of its three inputs are at 1.

3.7   State what is meant by a diode matrix. Draw a diode matrix to provide the outputs listed below when inputs A, B and C and their complements are applied.

$$ABC, AB\bar{C}, A\bar{B}C, \bar{A}BC, AB\bar{C}, \bar{A}B\bar{C}, \bar{A}\bar{B}C$$

3.8   Design a logic circuit to have the following characteristics: inputs A, B and C representing, in binary, the denary numbers 0 through to 7, where A is the least significant bit. There should be two outputs X and Y; X should be at 1 only when the input is an even number but not zero; Y should be at 1 only when the input is an odd number. Draw your circuit using either NAND or NOR gates only.

Fig. 3.15 Diode matrix decimal-to-bcd converter

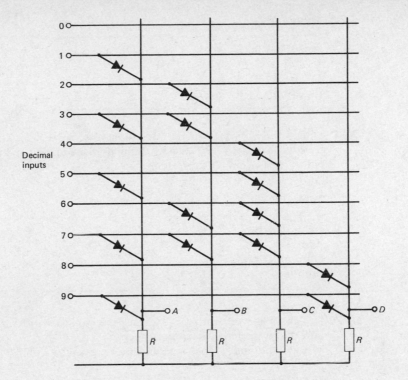

Decimal inputs

Fig. 3.16 Diode matrix octal decoder

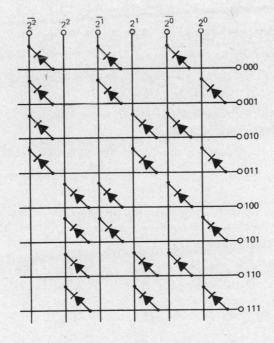

Table 3.7

A	0	0	0	0	1	1	1	1
B	0	0	1	1	0	0	1	1
C	0	1	0	1	0	1	0	1
F	1	1	1	0	1	0	0	0

3.9 Obtain the Boolean equation describing the logical operation of the circuit whose truth table is given by Table 3.7. Simplify the equation and then implement it using either NAND or NOR gates only.

Table 3.8

Input denary numbers	0	1	2	3	4	5	6	7
Output denary numbers	0	6	4	7	0	3	7	7

3.10 Design a logic circuit that will change the input denary numbers given in Table 3.8 into the given output denary numbers.

Both input and output numbers are represented in binary code.

Short Exercises

3.11 Show that the circuit given in Fig. 3.4 does produce the sum and carry outputs of a full-adder.

3.12 Draw a diode matrix suitable for converting 4-bit binary into hexadecimal.

3.13 Describe the operation of the diode matrix of Fig. 3.16.

3.14 Write down the function table for a 3-to-8 line demultiplexer.

3.15 Show that the circuit given in Fig. 3.12 generates the function $G_1 = \bar{A}B + A\bar{B}$, $G_2 = \bar{B}C + B\bar{C}$, $G_3 = \bar{C}D + C\bar{D}$, and $G_4 = D$.

3.16 Prove that equations (3.10) and (3.6) can be written in the form given, respectively, by equations (3.11) and (3.12).

3.17 Implement a binary half-adder using 4001 quad 2-input NOR gates only.

3.18 Show how two 7483 4-bit binary adders could be connected to act as a 6-bit binary adder.

3.19 Show how the 4077 quad exclusive-OR gate and a 4012 dual 4-input NAND gate can be connected as a magnitude comparator. Is the condition $A = B$ indicated by a low or a high output?

3.20 Repeat Exercise 3.19 using a 7451 dual 2-wide 2-input AOI gate instead of the 4077.

3.21 Show that the bcd-to-decimal converter of Fig. 3.11 could be implemented using the following ics: two 4011, two 4023, and one 4069.

3.22 Interconnect two ttl circuits 7483 to make a 6-bit full-adder.

3.23 Draw the diagram of a full-adder using NOR gates only.

3.24 Redraw Fig. 3.11 using 4011 and 4012 ics.

3.25 Implement the decimal-to-bcd converter of Fig. 3.10 using NOR gates only. Show the necessary connections when 7402 gates are used.

3.26 Derive the simpler version of the bcd-to-decimal converter mentioned on p. 73.

4 Bistable Multivibrators

The **bistable multivibrator**, or **flip-flop**, is a circuit that has two stable states. It is able to remain in either state for an indefinite period of time. The flip-flop will change state only when a trigger pulse is applied to one of its input terminals. Once the circuit has changed state, it will remain in its new state until another trigger pulse is applied to the circuit to make it revert back to its original state. A bistable multivibrator has two output terminals that are usually labelled as Q and $\bar{Q}$ since the logical state of one output terminal is *always* the complement of the logical state of the other terminal. When the circuit is in the state $Q = 1$, $\bar{Q} = 0$ it is said to be SET; conversely, when $Q = 0$ and $\bar{Q} = 1$ the circuit is said to be RESET.

Four types of flip-flop are available, known respectively as the S-R, the J-K, the D, and the T flip-flops. The S-R and J-K flip-flops can act as 1-bit stores and may be interconnected to provide various other more complex functions, such as counters and registers. The D flip-flop is used mainly to provide a time delay equal to the periodic time of the *clock* waveform. It can, however, also be connected to form counter circuits, etc. Finally, the T flip-flop acts as a *toggle*, that is it changes state every time the clock waveform is at logical 1.

The various kinds of flip-flop *can* be constructed by the suitable interconnection of two or more NOR or NAND gates, but it is more often convenient to use one of the integrated circuit versions that are readily available in the various logic families.

The S-R Flip-Flop

The **S-R flip-flop** has two input terminals labelled as S (for Set) and R (for Reset) and two output terminals labelled as Q and $\bar{Q}$. An S-R flip-flop may be operated either asynchronously or synchronously (i.e. clocked). When the circuit is not clocked it is often known as a **latch**.

The symbol for an S-R flip-flop is given in Fig. 4.1 and its truth table by Table 4.1. In the table the symbol Q represents the *present* state of the Q output terminal, and Q^+ represents the *next* state of that terminal *after* a set (S) or reset (R) pulse has been applied to the appropriate input terminal.

If the inputs S and R are both at logical 0, the output state of the circuit will not change and the flip-flop can be said to have stored one bit of information.

Table 4.1 S-R flip-flop truth table

S	R	Q	Q^+
0	0	0	0
0	0	1	1
1	0	0	1
1	0	1	1
0	1	0	0
0	1	1	0
1	1	0	X
1	1	1	X

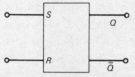

Fig. 4.1 Symbol for an S-R flip-flop

When $S = 1$ and $R = 0$, the flip-flop will change state if $Q = 0$ to $Q^+ = 1$ but it will not change state if $Q = 1$. Conversely, if $Q = 1$, a reset pulse, i.e. $R = 1$, $S = 0$, will cause the circuit to switch to $Q^+ = 0$, but the $R = 1$, $S = 0$ input condition will have no effect on the circuit if $Q = 0$ and $\bar{Q} = 1$.

Thus, $S = 1$, $R = 0$ will always produce the state $Q^+ = 1$; and $S = 0$, $R = 1$ will always give $Q^+ = 0$ regardless of the original state of Q.

Lastly, if $R = S = 1$ the flip-flop may or may not change state and the operation of the circuit is said to be *indeterminate*. This condition is denoted in the table by X.

The entries in the truth table can be used to derive the **state table** of the S-R flip-flop. The state table provides an alternative method of representing the operation of the flip-flop. The rows in the state table represent the output states of the circuit, while the columns correspond to the possible input states. An entry in a particular position in the state table represents the *next* state Q^+ of the output of the flip-flop for the circuit change, or *transition*, caused by applying the input shown by that column when the circuit is in the state represented by that row.

The state table of an S-R flip-flop is

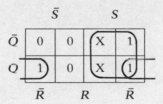

From the state table the **characteristic equations** of the S-R flip-flop can be written down:

$$Q^+ = S + Q\bar{R} \tag{4.1}$$

$$\overline{Q^+} = R + \bar{Q}\bar{S} \tag{4.2}$$

The characteristic equations can be used to derive the NOR and the NAND gate implementations of the S-R flip-flop.

The **transition table** of a flip-flop specifies the required inputs for every possible combination of present and next states of the flip-flop. The transition table of an S-R flip-flop is given by Table 4.2.

Table 4.2 S-R flip-flop transition table

Present state Q	Desired next state Q^+	Required inputs	
		S	R
0	0	0	X
0	1	1	0
1	0	0	1
1	1	X	0

The NOR gate implementation of the characteristic equations is shown in Fig. 4.2a. This circuit can be simplified by the removal of the redundant gates to produce the arrangement shown in Fig. 4.2b. The two outputs of the circuit are $\overline{Q\bar{R}+S}$ and $\overline{Q\bar{S}+R}$ respectively or, from equations (4.1) and (4.2), they are $\bar{Q}$ and Q. If now it is realised that the outputs of the left-hand gates, i.e. $\overline{R+\bar{Q}}$ and $\overline{S+Q}$, are equal to Q and $\bar{Q}$ respectively, i.e.

$$\overline{R+\bar{Q}} = \overline{R+R+\bar{Q}\bar{S}} = \overline{R+\bar{Q}\bar{S}} = \overline{\bar{Q}} = Q$$

the circuit can be still further reduced to give Fig. 4.2c.

The NAND gate version of the S-R flip-flop can be similarly determined with the result shown in Fig. 4.3a. The output of the upper NAND gate is $\overline{\bar{S}\bar{Q}} = S+Q$ and this is applied to one of the inputs of the lower NAND gate. The output of this gate is

$$\overline{R(S+Q)} = \overline{R+\overline{S+Q}} = \overline{R+\bar{S}\bar{Q}} = \overline{Q^+}$$

Similarly, this output can be written as $\overline{\bar{R}Q} = R+\bar{Q}$ and this is applied to the upper gate to produce an output of

$$\overline{\bar{S}(R+\bar{Q})} = S+\overline{R+\bar{Q}} = S+\bar{R}Q = Q^+$$

If the NAND gate S-R flip-flop is constructed using only two gates, as in Fig. 4.3b, the output of the upper gate is $\overline{R\bar{Q}} = \bar{R}+Q$ and so the output of the lower gate is

$$\overline{S(\bar{R}+Q)} = \bar{S}+\overline{\bar{R}+Q} = \bar{S}+R\bar{Q}$$

Therefore

$$\overline{Q^+} = \bar{S}+R\bar{Q} \tag{4.3}$$

The output of the lower gate is $\overline{SQ} = \bar{S}+\bar{Q}$ and so the output of the upper gate is

$$\overline{R(\bar{S}+\bar{Q})} = \bar{R}+\overline{\bar{S}+\bar{Q}} = \bar{R}+SQ$$

Therefore

$$Q^+ = \bar{R}+SQ \tag{4.4}$$

Comparing equations (4.3) and (4.4) with equations (4.1) and (4.2) it can be seen that the S and R signals are inverted. Equations (4.3) and (4.4)

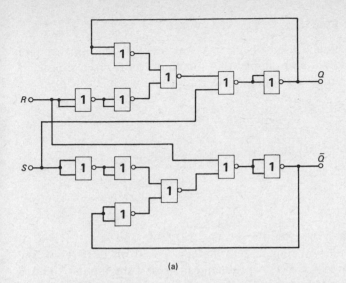
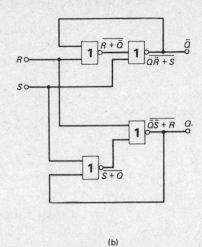

(a)

(b)

Fig. 4.2 Steps in the derivation of a circuit for the NOR gate version of an S-R flip-flop

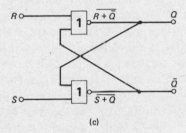

(c)

Fig. 4.3 (*a*) NAND gate S-R flip-flop, (*b*) alternative S-R flip-flop in which the indeterminate state is $S=R=0$

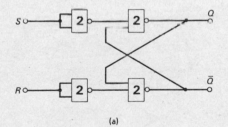

(a)

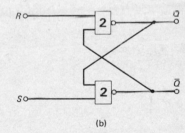

(b)

Table 4.3

S	0	0	1	1	0	0	1	1
R	0	0	0	0	1	1	1	1
Q	0	1	0	1	0	1	0	1
Q^+	X	X	1	1	0	0	0	1

relate to the truth table given by Table 4.3. This table represents a circuit whose indeterminate state is $S=R=0$ and for which the input conditions $S=R=1$ produce no change in the state of the circuit.

The Gated S-R Flip-Flop

A flip-flop is not generally used on its own but as part of a much larger digital system. Suppose, as an example, that the input to the reset terminal is provided by the output of a 2-input AND gate as shown in Fig. 4.4.

Fig. 4.4 S-R flip-flop with R input gated

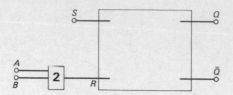

Consider a time when the circuit is set, $Q = 1$, $\bar{Q} = 0$, and $A = 1$, $B = 0$, so that $R = 0$. The circuit will remain set whether S is 0 or 1. If $S = 0$, the circuit will reset when $A = B = 1$. A problem can arise if the inputs A and B are supposed to change their states simultaneously to the state $A = 0$, $B = 1$; this would ensure that R remains at 0 so that the flip-flop remains set. It is difficult, however, to ensure that A and B change state simultaneously and the possibility exists that B may change state to 1 *before* A changes state from 1 to 0. Should this situation occur, R will be at 1 for a short period of time and the flip-flop will reset, $Q = 0$, $\bar{Q} = 1$, and will remain reset when A changes to 0.

To overcome this kind of difficulty a **clocked** or **gated S-R flip-flop** can be used. With such a circuit the S and R input states determine the final state of the flip-flop *but* the time at which any change in state takes place is determined by the clock.

Fig. 4.5 Clocked S-R flip-flop

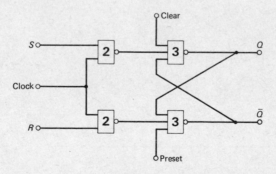

Fig. 4.5 shows how a NAND gate S-R flip-flop can be clocked or gated. Whenever the clock is at logical 0, the output of both the input NAND gates must be at logical 1 whatever the logical states of the S and the R inputs. Suppose the circuit is set; then $Q = 1$, $\bar{Q} = 0$ and so the upper right-hand gate has one input at 1 ($\bar{C}S + \bar{C}\bar{S}$) and one at 0 ($\bar{Q}$), and hence the Q output remains at 1. The lower right-hand gate has both of its inputs at 1 and so the $\bar{Q}$ output remains at 0. Only when the clock input is 1 can the appropriate

input gate (S or $R = 1$) have an input at 0 for a possible switching action to be initiated. This means that, when the clock is 1, the circuit operates in the same way as the NAND circuit of Fig. 4.3a.

The clocked S-R flip-flop is often provided with clear and preset terminals (Fig. 4.5) that allow the normal inputs to be overridden.

Switch Debouncing

When an electrical switch is operated it will generally bounce off its new contact several times (without retouching the original contact) before it settles down at its new position. In many cases such *contact bounce* is unimportant but when its presence is undesirable it can be eliminated by the use of an S-R flip-flop.

Fig. 4.6 Switch contact debouncing: (*a*) basic circuit, (*b*) S-R flip-flop connected as a debouncer

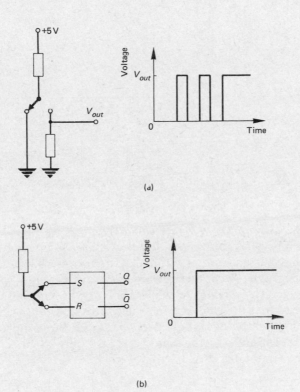

A circuit needing debouncing is shown in Fig. 4.6. When the switch operates, contact bounce produces a few blips of output voltage before a steady voltage is obtained. With the S-R flip-flop debouncing circuit of Fig. 4.6b the operation of the switch sets the flip-flop immediately. Its S terminal is taken high and the output voltage goes to, and remains at, +5 V.

(*Note*: a Schmitt trigger can also be used as a contact debouncer.)

The J-K Flip-Flop

Very often the indeterminate state $S = R = 1$ (or 0) of an S-R flip-flop cannot be permitted to occur and then an alternative circuit, known as the **J-K flip-flop**, is used. The operational difference between the S-R and J-K flip-flops lies in the response of the circuits to the input state $S = R = 1$. The truth table of a J-K flip-flop is shown by Table 4.4 and comparing this with the truth table of an S-R flip-flop (Table 4.1) makes it clear the J-K flip-flop *always* changes state when $J = K = 1$. Thus the J-K flip-flop **toggles** each time a clock pulse occurs.

Table 4.4 J-K flip-flop truth table

J	K	Q	Q^+
0	0	0	0
0	0	1	1
1	0	0	1
1	0	1	1
0	1	0	0
0	1	1	0
1	1	0	1
1	1	1	0

Fig. 4.7 Symbols for (*a*) a clocked J-K flip-flop, (*b*) a gated J-K flip-flop

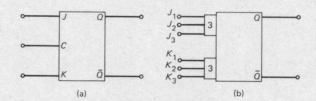

(a)   (b)

The symbol for a J-K flip-flop is given by Fig. 4.7*a*. A clock input is shown because this type of flip-flop is usually operated synchronously. Very often, clear and preset terminals are also provided. In addition, some J-K flip-flops are *gated*, that is they are provided with AND gates inside the ic package. The AND gate symbols are drawn touching the flip-flop symbol to indicate that the gates are internally provided (Fig. 4.7*b*). The J input of the flip-flop will be at logical 1 only when *all* three inputs J_1, J_2 and J_3 are at logical 1. Similarly, the K input is at 1 only when $K_1 = K_2 = K_3 = 1$.

The state table of a J-K flip-flop is

	$\bar{J}$		J	
$\bar{Q}$	0	0	1	1
Q	1	0	0	1
	$\bar{K}$	K	$\bar{K}$	

and is derived from the truth table. From the state table,

$$Q^+ = J\bar{Q} + \bar{K}Q \qquad\qquad (4.5)$$

$$\overline{Q^+} = \bar{J}\bar{Q} + KQ \qquad\qquad (4.6)$$

Table 4.5 J-K flip-flop transition table

Present state Q	Desired next state Q^+	Required inputs J	K
0	0	0	X
0	1	1	X
1	0	X	1
1	1	X	0

The transition table of a J-K flip-flop is given by Table 4.5.

Comparing equations (4.5) and (4.6) with equations (4.2) and (4.1) it is clear that the equations differ in form only in that the J input must be ANDed with $\bar{Q}$, and the K input must be ANDed with Q. Hence the J-K flip-flop can be implemented by the circuit of Fig. 4.8a. If the S-R flip-flop section of the circuit and the AND gates are implemented using NAND gates, Fig. 4.8b is obtained.

Fig. 4.8 J-K flip-flop

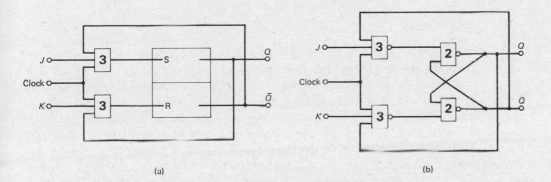

(a)   (b)

Unfortunately, in the J-K flip-flop shown in Fig. 4.8b there is a high probability of a race condition being set up when the input state is $J = K = 1$. Suppose, for example, that $Q = 0$, $\bar{Q} = 1$. When the clock changes from 0 to 1, the upper input NAND gate will be enabled and the flip-flop will set to give $Q = 1$, $\bar{Q} = 0$. Now the lower input NAND gate is enabled and the circuit resets. This gives $\bar{Q} = 1$ again and so the upper gate is enabled once more and the circuit sets again and so on for as long as the clock remains at 1. When the clock goes to 0, the circuit will remain in whichever state it happens to be in at that moment, and this is unpredictable.

Clearly, an oscillation of the output state of a flip-flop cannot be tolerated. One method of overcoming this problem is to arrange that the circuit only operates at the negative-going edges of the clock pulses; another method is to use a master-slave arrangement.

With both these types of J-K flip-flop the data present at the input terminals just prior to a clock edge determines the output Q^+ after the clock has changed state.

Master-Slave J-K Flip-Flop

The circuit of a **master-slave J-K flip-flop** is shown by Fig. 4.9. Suppose that initially the circuit is set so that $Q = 1$ and $\bar{Q} = 0$ and $J = K = 0$. When the clock is 0, the master flip-flop will have both inputs at 0 and so its outputs, labelled as Q' and $\bar{Q}'$ will remain unchanged at $Q' = 1$, $\bar{Q}' = 0$. The clock input to the slave flip-flop is inverted and so this part of the circuit has $J = 1$, $K = 0$. This is the set condition and so the output is unchanged.

Fig. 4.9 Master-slave J-K flip-flop

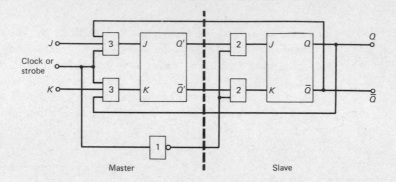

When the clock is 1, there will be no change in state *unless* $J = 0$, $K = 1$ or $J = K = 1$. In either case the three inputs to the lower AND gate are all at 1 and so the master flip-flop has inputs $J = 0$, $K = 1$. The master resets to have $Q' = 0$, $\bar{Q}' = 1$. The clock input to the slave is inverted, i.e. at 0, and so both inputs to the slave remain at 0. Hence the slave does not *change* state. At the end of the clock pulse clock becomes 1 and now the slave flip-flop has $J = 0$, $K = 1$ and resets to the state $Q = 0$, $\bar{Q} = 1$. Thus the circuit does *not* change state until the end of the clock pulse. The race condition previously described cannot now occur since, when $J = \bar{Q} = 1$, the clock *must* be at 0.

Now consider the operation of the circuit when it is initially in the reset state, i.e. $Q = 0$, $\bar{Q} = 1$. When $J = 1$, $K = 0$ or $J = K = 1$, all three inputs to the upper AND gate will be at 1 immediately the clock becomes 1. The J input of the master flip-flop is 1 and so this part of the circuit sets to give $Q' = 1$, $\bar{Q}' = 0$. As before this state cannot be passed onto the slave flip-flop because the inverted clock inhibits the slave AND gates. When the clock changes from 1 to 0, the J input of the slave becomes 1 and the slave sets; now $Q = 1$, $\bar{Q} = 0$. Again, there is no risk of the circuit oscillating because when the output Q becomes 1 the clock is 0.

Another version of the master-slave J-K flip-flop is shown in Fig. 4.10. Suppose that the circuit is set with $Q = 1$, $\bar{Q} = 0$ and that $J = K = 1$. When the clock is at 0, the logic states at various points in the circuit are: Gates A and B both have logic 1 at their outputs and, since the master flip-flop must also be set, gate E has both its inputs at 1 and so its output is 0. Therefore, gate G has both inputs at 0 confirming that $Q = 1$. Gate F has one input at 1 and the other input at 0 so that its output is 1; thus the output of gate H is 0.

Fig. 4.10 An alternative master-slave J-K flip-flop

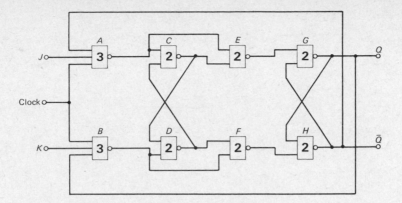

Fig. 4.10 An alternative master-slave J-K flip-flop

When the clock goes to 1, the output of gate B becomes 0. The two inputs to gate D are now at 1 and 0 so that its output goes to 1. This means that gate C has both inputs at 1 and its output switches to 0. The master flip-flop is now reset. Both gates E and F have one input at 1 and one input at 0 so that their outputs are 1. Now gate g has one input at 1 and the other at 0 so that its output O *remains* at 1. Similarly, gate H has both inputs at 1 and so Q remains at 0. This means that the logic state of the master is *not* passed on to the slave.

When the clock pulse ends, the output of gate B goes to logical 1 and then gate D has inputs at 0 and 1 and so its output remains unchanged at 1. This means that gate F now has both of its inputs at 1 and so its output becomes 0. Now gate H has inputs at 0 and 1 and its output Q becomes 1. In turn, this makes both of the inputs of gate G become logical 1 and so the output Q of this gate becomes 0. The slave has now reset.

The operation of the circuit for the conditions $Q = 0$, $\bar{Q} = 1$ and $J = 1$, $K = 0$ and for $J = K = 1$ is very similar (Exercise 4.21).

For any master-slave flip-flop the J and K inputs must be held constant while the clock is at logical 1.

Edge-triggered J-K Flip-Flops

The **edge-triggered J-K flip-flop** behaves in exactly the same way as a master-slave circuit but its internal logic circuitry is different. The transfer of data to the output terminals may take place on either the negative or the positive edges of the clock waveform. The symbol for a J-K flip-flop is marked with a small wedge drawn at the clock input to indicate that the circuit is edge-triggered.

In general, standard ttl flip-flops are master-slave devices, and low-power Schottky flip-flops are edge-triggered, e.g. 7473 is master-slave, 74LS73 is edge-triggered, 7476 is master-slave, 74LS76 is edge-triggered.

The input data must be held constant for some time both before and after the clock transition takes place to ensure that it will be transferred to the output. The **set-up time** is the time that elapses between the leading edge of

Fig. 4.11 Set-up time
and hold-up time for
an edge-triggered
flip-flop

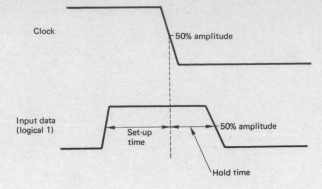

an input data pulse and the triggering edge of the clock pulse. Set-up time is
illustrated for a negative-edge-triggered flip-flop by Fig. 4.11.

The **hold-up time** of an edge-triggered flip-flop is the interval of time
between the clock pulse transition changing the state of the output and the
end of the input data pulse. The hold time is illustrated by Fig. 4.11. After
the hold time interval has passed the J and K input data may be changed
without affecting the output.

Typical figures for a low-power Schottky circuit are a set-up time of 20 ns
and a hold time of very nearly zero.

Clear and Preset Controls

The operation of a flip-flop is determined by the states of its input terminals,
S-R or J-K, when a clock pulse is applied. The *initial* state of the circuit is
not, however, so determined and may have either of its two possible values.

If Clear and Preset controls are added they can be employed to specify
the initial state of the circuit. The operation of these controls is shown by
the truth table of a J-K flip-flop given in Table 4.6.

It should be noted that the Preset and Clear terminals when at logical 0
set or reset the output Q, *whatever* the states of the clock or the J or K

Table 4.6

Preset	Clear	Clock	J	K	Q	Q⁺
0	1	X	X	X	0	1
0	1	X	X	X	1	1
1	0	X	X	X	0	0
1	0	X	X	X	1	0
1	1	←——— as for J-K ———→				

Fig. 4.12 The J-K flip-flop as a divide-by-two circuit

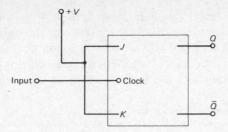

inputs. If both the Preset and Clear inputs are at logical 1, the operation of the circuit will be determined in a manner already described. If the Preset and/or Clear facilities are not required, the terminals must be connected to a high voltage level to disable their action.

The J-K flip-flop can be used as a **divide-by two** circuit by connecting it in the manner shown in Fig. 4.12. With the J and K input terminals connected to +V, both inputs are held in the logical 1 state and the circuit *toggles* with each clock pulse. The toggle action may, of course, occur at either the leading or the trailing edge of the clock pulse, depending upon the device concerned.

The relative merits of the two kinds of J-K flip-flop are as follows. For many master-slave circuits the J and K inputs must not be allowed to change while the clock is at logical 1. This is because any input that changes from 0 to 1 during this time will produce an effect known as "ones catching". This means that, if either the J or the K input or both should momentarily change from 1 to 0 while the slave is enabled, the circuit *may* change state even though the input data has restored to its original value by the time that the clock transition occurs. Thus the circuit may change state when it should not do so. The problem can always be overcome either by making sure that both the J and K inputs remain constant while the clock is 1, or by using short clock pulses. However, it is probably better to use edge-triggered devices instead. The data inputs to an edge-triggered device can be changed when the clock is at either 1 or 0. The new output state will be determined by the states of the J and K inputs at the start of the set-up time (during which they must, of course, be kept constant). Because the input data is not first stored by a master and then transferred to a slave, the edge-triggered flip-flop is faster to operate.

CMOS Flip-Flops

Flip-flops *can* be constructed using cmos NOR or NAND gates in the ways previously discussed but cmos flip-flops do not use this approach. This is because transmission gates (p. 52) are easy to make using cmos techniques and their use allows different, and simpler, circuitry to be used.

The D Flip-Flop

The **D flip-flop** has a single D input terminal. Its logical operation is such that its Q output terminal *always* takes up the same logic state as its D input. The truth table of a D flip-flop is given by Table 4.7. Any change in state occurs when the clock is 1. When the clock is 0 the output is unable to follow the D input and the circuit is said to be *latched*.

Table 4.7 D flip-flop truth table

Clock	D	Q	Q⁺
1	0	0	0
1	0	1	0
1	1	0	1
1	1	1	1

The state table of a D flip-flop is

From the state table the characteristic equation of the circuit can be written as

$$Q^+ = D \tag{4.7}$$
$$\overline{Q^+} = \bar{D} \tag{4.8}$$

Table 4.8 gives the transition table of the D flip-flop. Comparing it with the characteristic equation and the transition table of the J-K, or the S-R flip-flop (equations 4.1, 4.2, 4.5, and 4.6), it can be seen that the D flip-flop can be produced by a J-K or S-R flip-flop having complementary inputs.

Table 4.8 D flip-flop transition table

Present state Q	Desired next state Q⁺	Required input
0	0	0
0	1	1
1	0	0
1	1	1

Complementary input signals can easily be arranged by connecting an inverter between the D input and the K terminal as shown by Fig. 4.13. There will not be an indeterminate state if an S-R flip-flop is used since the condition $S = R = 1$ is not possible.

The main use for the D flip-flop is as a delay circuit; the output takes up the same logical state as the input after a time delay equal to the period of

Fig. 4.13 D flip-flop

Fig. 4.14 Use of a D flip-flop to divide-by-two

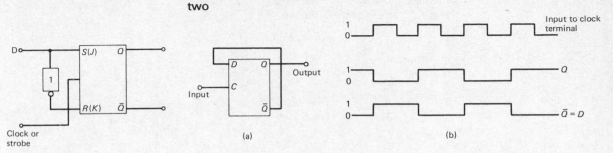

(a)

(b)

Clock or strobe

Input

one clock pulse. Some D flip-flops are manufactured in master-slave versions and may also be provided with clear and preset controls.

A variation of the circuit is known as the **D latch** in which, when the clock is 1, the output immediately takes up the input state. It then latches in that state until the next clock pulse arrives. In other words, the D latch is able to change its output at any time while the clock remains at 1.

Both D latches and flip-flops are available in both the ttl and cmos families.

A D flip-flop can be used as a divide-by-two circuit by merely connecting its Q and D terminals together (Fig. 4.14).

The block diagram of a ttl D flip-flop is given by Fig. 4.15 (minus the clear and preset controls). Analysis of the circuit operation will show that the output Q changes to the state of the D input *only* at the positive edge of the clock pulse.

Fig. 4.15 Construction of a D flip-flop

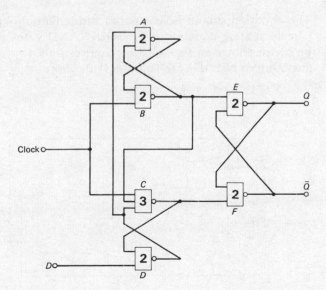

Table 4.9

Present state	Desired next state	Required inputs		
Q	Q^+	J	K	D
0	0	0	0	0
0	0	0	1	0
0	1	1	0	1
0	1	1	1	1
1	1	0	0	1
1	0	0	1	0
1	1	1	0	1
1	0	1	1	0

It has already been shown that a D flip-flop can be made from a J-K flip-flop but the reverse process is also possible. The transition table of such a circuit is given by Table 4.9. From this table the state table can be written down, mapping this time for D instead of for Q^+ (although really the same thing). Thus

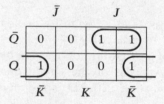

From the state table,

$$D = J\bar{Q} + \bar{K}Q \tag{4.9}$$

(This equation should be compared with equation 4.5.)

Implementing equation (4.9) using NAND gates and a D flip-flop gives the circuit shown in Fig. 4.16; it requires only two integrated circuits: one quad 2-input NAND gate and one D flip-flop.

Fig. 4.16 D flip-flop connected as a J-K flip-flop

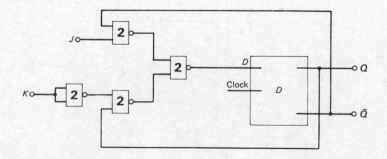

The T Flip-Flop

The fourth type of flip-flop is known as the **T flip-flop** or *trigger* flip-flop. Its truth table is given in Table 4.10. It can be seen that when the clock is 1 the Q output of the circuit will change state, or toggle, each time a pulse is applied to the trigger (T) terminal. The transition table of a T flip-flop is given by Table 4.11. Note that the columns of Tables 4.10 and 4.11 are the same. The state table is

	$\bar{T}$	T
$\bar{Q}$	0	1
Q	1	0

and from this the characteristic equations of the T flip-flop can be written as

$$Q^+ = T\bar{Q} + \bar{T}Q \qquad\qquad (4.10)$$
$$\overline{Q^+} = \bar{T}\bar{Q} + TQ \qquad\qquad (4.11)$$

Comparing these equations with those for the J-K flip-flop (4.5 and 4.6), it is clear that the J-K circuit can be made to act as a T flip-flop by connecting its J and K terminals together, so that at all times $J = K$ and using the common point as the T input (see Fig. 4.17).

Fig. 4.17 T flip-flop

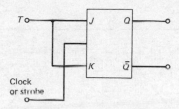

Table 4.10 T flip-flop truth table

Clock	T	Q	Q^+
1	0	0	0
1	0	1	1
1	1	0	1
1	1	1	0

Table 4.11 T flip-flop transition table

Present state Q	Desired next state Q^+	Required input T
0	0	0
0	1	1
1	0	1
1	1	0

ECL and I²L Flip-Flops

Flip-flops are also manufactured using the ecl and I²L techniques, although the number of ecl devices available is very limited compared with ttl and cmos, whilst I²L is exclusively an lsi technology. Their relative merits have been discussed in Chapter 2.

Exercises 4

4.1 With the aid of a diagram explain the operation of an S-R flip-flop. List the disadvantages of the S-R flip-flop and explain how these are overcome by the master-slave J-K flip-flop.

4.2 With the aid of a logic diagram show how two NOR gates can be connected to act as a S-R flip-flop. Write down the truth table of the circuit. Explain why two NAND gates similarly connected differ in their logic operation. How can the NAND gate version of the S-R flip-flop be modified to produce the basic S-R flip-flop operation?

4.3 Use logic diagrams to explain the operation of both a D flip-flop and a J-K flip-flop. Explain how a) a D flip-flop can be made from a J-K flip-flop, b) how a J-K flip-flop can be made from a D flip-flop.

4.4 Analyse the operation of the D flip-flop given in Fig. 4.15.

4.5 Explain the operation of the circuit shown in Fig. 4.18.

Fig. 4.18

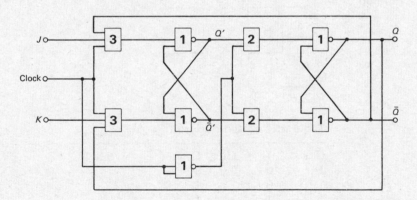

4.6 Write down the truth table of an S-R flip-flop and use it to obtain a) the state table, b) the characteristic equation, c) the transition table of the circuit. Use your results to derive *either* the NOR *or* the NAND gate version of the circuit.

4.7 Write down the truth tables for the J-K, D and T flip-flops. Hence obtain the characteristic equation of each circuit. Compare your results and hence explain how the J-K circuit can be operated as a) a D flip-flop, b) a T flip-flop.

4.8 Briefly explain the principle of operation of master-slave and edge-triggered J-K flip-flops. Most low-power Schottky versions of J-K flip-flops are edge-triggered devices. Suggest some reasons for this.

Short Exercises

4.9   Show, by means of a transition table, that two cross-coupled NAND gates can store data.

4.10   With the aid of truth tables show that all the functions of the D, T and S-R flip-flops are included in the J-K flip-flop.

4.11   Draw the circuit of an S-R flip-flop that uses NOR gates only. Modify the circuit to produce a D flip-flop. In both cases give the truth table of the circuit.

4.12   Explain how a D flip-flop can be used as a divide-by-two circuit. Give the truth table of the divider.

4.13   Draw the circuit, and explain the operation of, a master-slave J-K flip-flop.

4.14   Draw the logic circuit of either a NAND or a NOR gate clocked latch. By means of a truth table show how the Q and $\bar{Q}$ outputs respond to signals applied to the S and R inputs. Hence state the basic disadvantage of the latch.

4.15   Draw a diagram to show how a J-K flip-flop may be used as a D flip-flop. Give the truth table.

4.16   Explain what is meant by the master-slave operation of a flip-flop. Why are such circuits usually negative-edge-triggered devices?

4.17   Draw a diagram to show how a J-K flip-flop can be derived from an S-R circuit. Also show how the J-K flip-flop can be used to divide-by-two.

4.18   Explain the operation of the clocked S-R flip-flop shown in Fig. 4.5.

4.19   Prove that the outputs of the circuit given in Fig. 4.8b are $Q^{+} - J\bar{Q} + \bar{K}Q$, $\overline{Q^{+}} - \bar{J}\bar{Q} + KQ$.

4.20   Show how a J-K master-slave flip-flop can be fabricated using 8 NAND gates and one inverter.

4.21   Fig. 4.10 shows the block diagram of a ttl master-slave J-K flip-flop. Explain its operation when $J = K = 1$.

4.22   List the advantages of edge-triggered flip-flops over their master-slave equivalents.

4.23   Define the terms *set-up time* and *hold time* when applied to an edge-triggered flip-flop. Give typical values for a ttl device and explain the importance of these quantities.

4.24   Many flip-flop circuits are provided with clear and preset controls. Explain their functions.

5 Counters and Shift Registers

Introduction

A **counter** is a digital circuit that consists of a number n of D or J-K flip-flops connected in cascade, whose function is to count the number of pulses applied to an input terminal. The count may be indicated using the straightforward binary code, or binary-coded-decimal, or some other code. Alternatively, decoding circuitry may be employed to give a direct readout of the count. If only the Q output of the final flip-flop is employed the counter will act as a divide-by-n circuit. The maximum number of possible 1 and 0 states is known as the **modulus** of the counter and this cannot be greater than 2^n.

A counter can be constructed using a number of ic flip-flop packages, with some gating if the desired count is less than 2^n, or a number of msi counters are available in both the ttl and the cmos families. In either case the counter may be operated either synchronously or non-synchronously.

Counters find application in many kinds of digital equipments, such as for example, the direct counting of pulses, division-by-n, and frequency and time measurements.

A **shift register** also consists of a number of J-K flip-flops connected Q to J and $\bar{Q}$ to K throughout, and operates in a similar manner to a counter. Shift registers are used as temporary stores, serial-to-parallel converters (and vice versa), and can also be connected to operate as *ring counters* and sequence generators.

A description of the operation of the various counters and registers has been given in [DT&S]. In this chapter a simple design method will be introduced and further applications considered.

Non-synchronous Counters

A single J-K or D flip-flop can be connected to act as a divide-by-two circuit (Figs. 4.12 and 4.14). For counts greater than 2, a number n of flip-flops must be connected in cascade to give a maximum count of $2^n - 1$. Since the first count is 0 this means that n flip-flops can give a count of 2^n.

The idea is illustrated by Figs. 5.1a and b. Fig. 5.1a shows how two D flip-flops can be connected to give a maximum count of 3 (divide-by-4) and Fig. 5.1b shows three J-K flip-flops connected as a divide-by-8 circuit. Each stage is clocked at one-half the rate of the preceding stage and so the states of the Q outputs appear to *ripple* through the circuit. For this reason the circuit is often known as a **ripple counter**.

Fig. 5.1 (*a*) D flip-flops connected as a divide-by-4 circuit (*b*) J-K flip-flops connected as a divide-by-8 circuit

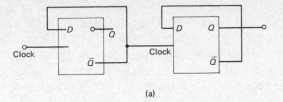

(a)

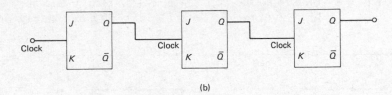

(b)

The main disadvantage of this type of circuit is that the propagation delays of the individual stages are additive and the total delay may severely limit the maximum possible frequency of operation.

Example 5.1

The flip-flops in a 4-bit counter each introduce a maximum delay of 40 ns. Calculate the maximum clock frequency.

Solution
Total propagation delay $= 4 \times 40 = 160$ ns.
Thus, maximum clock frequency $= 1/(160 \times 10^{-9}) = 6.25 \times 10^{6}$ (*Ans*)

Three-stage and four-stage (bit) ripple counters can also be designed using an extension of the method shown in Fig. 5.1. Fig. 5.2 shows a **4-bit counter** that uses two 7473 dual J-K master-slave flip-flops. The timing diagram of the 4-bit counter is given in Fig. 5.3 and its truth table is given by Table 5.1 and these can be used to explain the action of the circuit.

Table 5.1 4-bit counter truth table

Clock pulse	0	1	2	3	4	5	6	7	8	9	10	11	12	13	14	15	16
Q_A	0	1	0	1	0	1	0	1	0	1	0	1	0	1	0	1	0
Q_B	0	0	1	1	0	0	1	1	0	0	1	1	0	0	1	1	0
Q_C	0	0	0	0	1	1	1	1	0	0	0	0	1	1	1	1	0
Q_D	0	0	0	0	0	0	0	0	1	1	1	1	1	1	1	1	0

Suppose that, initially, the circuit has each of its four stages reset, i.e. $Q_A = Q_B = Q_C = Q_D = 0$. The trailing edge of the first clock pulse will cause the first flip-flop to toggle so that $Q_A = 1$. The second clock pulse will again toggle flip-flop 1 and the change of Q_A from 1 to 0 will toggle the second flip-flop. Now $Q_A = 0$, $Q_B = 1$. Clock pulse 3 will toggle the first flip-flop so that Q_A becomes 1 but the change in Q_A from 0 to 1 will have no effect on the logical state of flip-flop 2. Now $Q_A = Q_B = 1$ and $Q_C = Q_D = 0$. The fourth clock pulse causes Q_A to change from 1 to 0 and this change in state resets the second flip-flop. Q_B goes from 1 to 0 and in so doing causes the third flip-flop to toggle. Now, $Q_A = Q_B = Q_D = 0$ and $Q_C = 1$.

Fig. 5.2 4-bit ripple counter using two 7473 dual J-K flip-flops

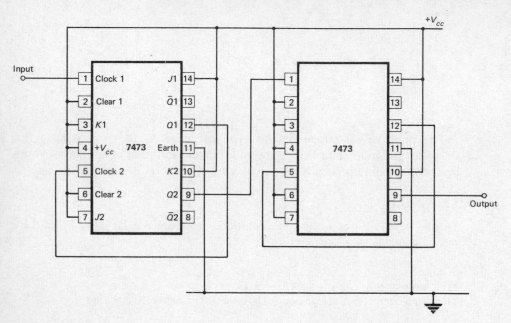

Fig 5.3 Timing diagram of a 4-bit ripple counter

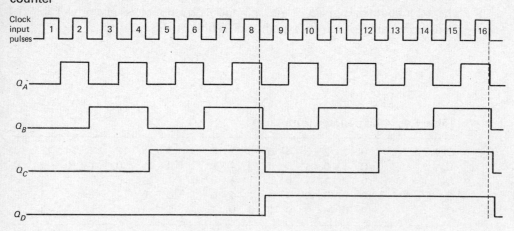

Further operation of the counter as succeeding clock pulses are applied to the circuit is summarized by its truth table and the timing diagram.

When more than two or, perhaps, three stages are required it is usually cheaper to use an msi counter. Examples of these include: 7493 4-bit counter, 74293 4-bit counter, 4024 7-bit counter, 4040 12-bit counter, and 4020 14-bit counter.

If the output is taken from the Q terminal of the final flip-flop, as in Figs. 5.1 and 5.2, the circuit acts as divider. The count at any time can be indicated by utilizing the Q outputs of each stage and some kind of indicating device, e.g. led's (see Chapter 6).

For some applications a binary readout is undesirable; perhaps the circuit is required to generate an output pulse when the count reaches some particular number. When this is the case a decoder must be added to the circuit. Decoding is easily achieved by connecting the Q output of each stage that is at 1, and the $\bar{Q}$ output of each stage that is at 0, for the required count to a NAND gate. The output of the NAND gate will be at logical 0 *only* when the count is at the required point. An example of the technique is shown in Fig. 5.4. The enabling clock pulse is not applied until all the flip-flops have switched and this ensures that false signals, or glitches, do not appear at the output.

A ripple counter will count down (instead of up) if the $\bar{Q}$ outputs of each stage are used as the clock signal for the following stage.

Fig. 5.4 Counter with decoded outputs

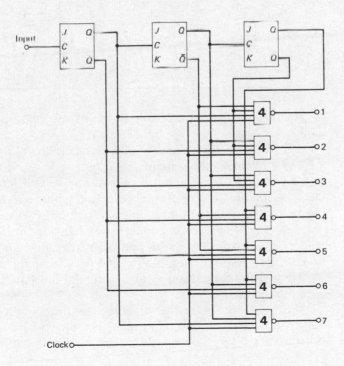

Reducing the Count to Less than 2^n

Very often a counter is required to have a count of less than 2^n, where n is the number of stages. A ripple counter must then be modified so that one or more of the possible counts are omitted. For example, if a count of 6 is required, a 3-stage counter must be used with *two* of its counts eliminated. The three ways in which this can be achieved are known as *a*) the feedback method, *b*) the reset method, *c*) the preset method, and have each been described previously [DT&S].

The design of the reset type of counter must start with a decision on the number n of stages that are necessary. A NAND gate is then required whose output is connected to all the reset, or clear, flip-flop inputs in parallel. The inputs to the NAND gate must consist of the Q output of each stage that is at logical 1 when the required count is reached.

1 *Decade counter* Four stages are necessary: for a count of 10 or 1010, then $Q_A = 0$, $Q_B = 1$, $Q_C = 0$, $Q_D = 1$, and hence the NAND gate inputs are Q_B and Q_D. The circuit is shown by Fig. 5.5.

2 *Divide-by-6 counter* Three stages are needed: for a count of 6 or 110, then $Q_A = 0$, $Q_B = 1$, $Q_C = 1$ and these are the necessary NAND gate inputs.

3 *Divide-by-5 counter* Three stages are needed: for a count of 5 or 0101, then $Q_A = 1$ and $Q_C = 1$ and so the NAND gate inputs are Q_A and Q_C.

Fig. 5.5 Decade counter

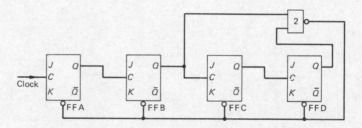

Ripple counters are available in the ttl and cmos logic families. Examples are

a) CMOS: 4020 14-stage, 4024 7-stage, 4040 12-stage ripple counters.

The 4020 and 4040 circuits have buffered output terminals (not Q_2 and Q_3 on the 4020), one input terminal, and a reset terminal, while the 4024 has seven buffered outputs and reset and input terminals. These devices can be made into divide-by-n counters using the reset method.

b) TTL: 7490 decade counter, 7492 divide-by-12 counter, 7493 4-bit binary ripple counter, 74196 presettable decade counter, 74197 presettable 4-bit ripple counter, and 74390 dual decade ripple counter.

All of these devices are available in both the standard and the low-power Schottky versions.

The 7493 consists of two sections: one a single divide-by-two circuit, the other three J-K flip-flops connected as a divide-by-8 ripple counter. For the circuit to act as a divide-by-16 counter the Q output of the first flip-flop

Fig 5.6 7493 4-bit binary counter

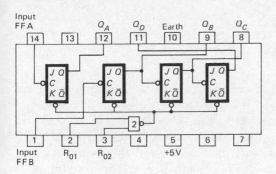

Table 5.2

Divide-by-n	R_{01} to:	R_{02} to:
$n = 7$	Q_A	$Q_B Q_C$
$n = 9$	Q_A	Q_D
$n = 10$	Q_B	Q_D
$n = 11$	Q_A	$Q_B Q_D$
$n = 12$	Q_C	Q_D
$n = 13$	Q_A	$Q_C Q_D$
$n = 14$	$Q_B Q_D$	Q_C
$n = 15$	Q_A	$Q_B Q_C Q_D$

must be connected to the clock input of the second flip-flap. Referring to Fig. 5.6, terminals 1 and 12 are linked together.

The 7493 can be modified, using the reset method, to give a count other than 2, 8 or 16. The circuit will reset at the required count if the two inputs that are then at 1 are fed back to the reset inputs R_{01} and R_{02}. Table 5.2 gives examples of the connections necessary for various division ratios.

The 7490 decade counter can also be modified to produce alternative counts. This device (Fig. 5.7) consists of four master-slave J-K flip-flops connected to provide a divide-by-2 counter and a divide-by-5 counter. It can be seen from the figure how the three flip-flops are connected to give a count of 5. In addition to the reset terminals R_{01} and R_{02}, two further terminals R_{91} and R_{92} are provided which allow the counter to be set to a count of 9. When this facility is not required these two terminals should be earthed.

Fig. 5.7 7490 decade counter

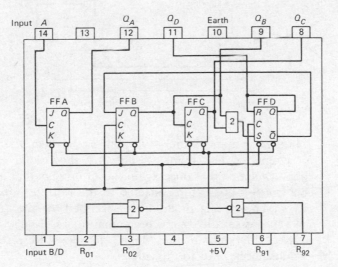

Fig. 5.8 Timing diagram for a divide-by-6 counter

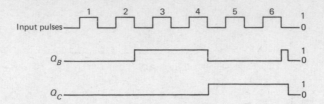

For operation as a bcd decade counter, the B/D input pin must be connected to the Q_A terminal but if a symmetrical divide-by-10 count is needed then Q_D should be connected to the A input. The input signal is then applied to the B/D input terminal and the output is taken from the Q_A terminal. As for the 7493 the outputs that are high at the required count must be connected to the R_{01} and R_{02} terminals, e.g. for a count of 6 connect Q_B to R_{01} and Q_C to R_{02}. The timing diagram for this particular case is given in Fig. 5.8. Q_B goes to logic 1 at the trailing edge of the second pulse and then goes back to 0 at the end of the fourth pulse. At this instant Q_C becomes 1. When the trailing edge of the sixth input pulse occurs, Q_B goes to 1 again and now both Q_B *and* Q_C are 1 and so the circuit resets its count to 0. Thus, Q_B is only at 1 for a very short period of time so that a voltage spike appears at the output terminal.

Fig. 5.9 7492 connected as a divide-by-9 counter

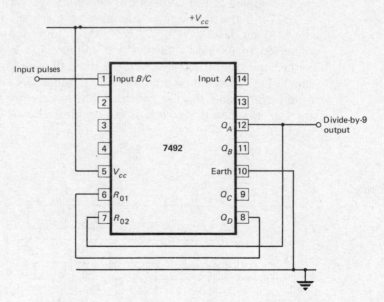

In similar manner the 7492 divide-by-12 counter can also be connected for counts of less than 12. Fig. 5.9 shows one example, a divide-by-9 counter. Notice that once again the Q outputs that are high at the required count of 9 (1001) are connected to the reset terminals.

The simple reset method, as described for the 7490, the 7492 and the 7493, possesses the disadvantage that the reset pulse only lasts for as long as it takes for one of the "high" flip-flops to reset. This can, in some circumstances, lead to difficulties.

Synchronous Counters

For many applications of counters the time taken for a clock pulse to ripple through the circuit and/or the glitches that often arise are not acceptable. The speed of operation can be considerably reduced if all the flip-flops are simultaneously clocked. This is known as **synchronous operation.**

The clock input of each flip-flop is directly connected to the clock line. The design of a synchronous counter when the count is required to be 2^n, where n is the number of stages is determined in the following manner.

The nth stage must change its state only for those clock pulses that arrive when all the preceding stages are set, i.e. have $Q = 1$.

The reason for this statement can be understood by considering the required values for Q_A, Q_B, etc. as the count increases. For example, for a count of 5, $Q_A = 1$, $Q_B = 0$, $Q_C = 1$ and $Q_D = 0$.

Suppose that a divide-by-8 counter is to be designed. Three stages are necessary. The first stage must toggle on each clock pulse and so $J_A = K_A = 1$. The second stage must toggle only when $Q_A = 1$ and to achieve this Q_A is connected to both J_B and K_B. Stage C must toggle only when *both* Q_A and Q_B are high and so these terminals are connected to the inputs of a 2-input AND gate. The output of the AND gate is then connected to both J_C and K_C. Thus the required circuit is shown by Fig. 5.10.

Fig. 5.10 Divide-by-8 synchronous counter

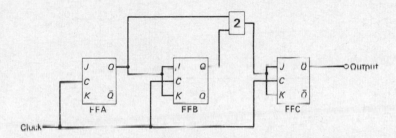

This principle is easily extended to obtain a 4-bit (divide-by-16) counter, a 5-bit (divide-by-32) counter, and so on, and Fig. 5.11 shows the circuit of the 4-bit counter. The operation of the 4-bit synchronous counter is summarized by its timing diagram, given by Fig. 5.12.

Fig. 5.11 Divide-by-16 synchronous counter

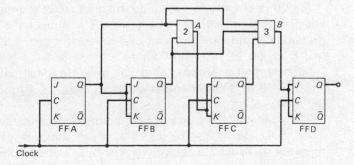

A divide-by-3 counter is easily obtained using two flip-flops with Q_A connected to J_B and $\bar{Q}_B$ connected to J_A, and also with $K_A = K_B = 1$ (see Fig. 5.13a). Flip-flop A is unable to toggle when $Q_B = 1$ and so the circuit counts to 3, as indicated by the timing diagram of Fig. 5.13b.

Fig. 5.12 Timing diagram for a 4-bit synchronous counter

Counts that are a multiple of 3 *or* the product of 3 and 2^n can also be easily obtained. Some examples are given in Figs. 5.14a and b.

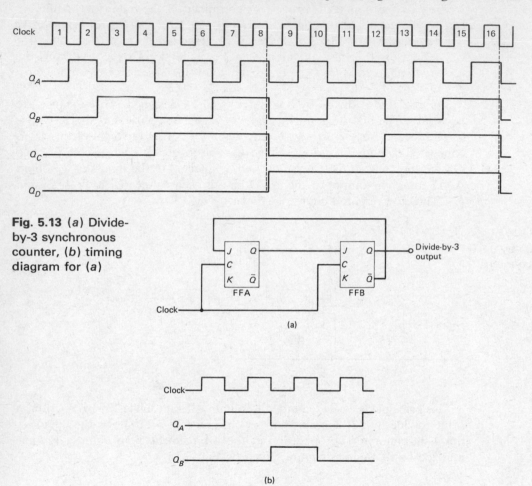

Fig. 5.13 (a) Divide-by-3 synchronous counter, (b) timing diagram for (a)

For other counts, and when an irregular count sequence is wanted, a more complex design method is necessary. The aim of the design method is to produce the minimal Boolean expressions for the gating required for the J and K inputs of each flip-flop.

Suppose that a decade counter is to be designed with the count progressing from 0 to 9 and then back to 0. A count of 10 requires the use of *four* flip-flops since $2^3 = 8$.

The state table for the decade counter must first be written down, see Table 5.4. The transition table of a J-K flip-flop given by Table 4.5 is repeated in Table 5.3 for convenience in understanding the state table.

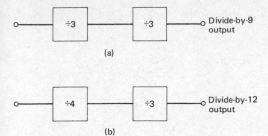

(a)

(b)

Divide-by-9 output

Divide-by-12 output

Fig. 5.14 Showing methods of obtaining counts of (*a*) a multiple of 3, (*b*) a product of 3 and 2^n

Table 5.3 J-K flip-flop transition table

Q	Q^+	J	K
0	0	0	X
0	1	1	X
1	0	X	1
1	1	X	0

Table 5.4 Decade counter state table

	Present state				Next state				Required J and K inputs							
	Q_D	Q_C	Q_B	Q_A	Q_D^+	Q_C^+	Q_B^+	Q_A^+	J_D	K_D	J_C	K_C	J_B	K_B	J_A	K_A
0	0	0	0	0	1	0	0	0	1	0	X	0	X	0	X	1	X
1	0	0	0	1	2	0	0	1	0	0	X	0	X	1	X	X	1
2	0	0	1	0	3	0	0	1	1	0	X	0	X	X	0	1	X
3	0	0	1	1	4	0	1	0	0	0	X	1	X	X	1	X	1
4	0	1	0	0	5	0	1	0	1	0	X	X	0	0	X	1	X
5	0	1	0	1	6	0	1	1	0	0	X	X	0	1	X	X	1
6	0	1	1	0	7	0	1	1	1	0	X	X	0	X	0	1	X
7	0	1	1	1	8	1	0	0	0	1	X	X	1	X	1	X	1
8	1	0	0	0	9	1	0	0	1	X	0	0	X	0	X	1	X
9	1	0	0	1	10	0	0	0	0	X	1	0	X	0	X	X	1
10	X	X	X	X	X	X	X	X	X	X	X	X	X	X	X	X	X
↓																
15	X	X	X	X		X	X	X	X	X	X	X	X	X	X	X	X

In Table 5.4 Q_A is the least significant output; the *present state* columns show the states of the four flip-flops for each count from 0 to 9 and the *next state* columns show the required changes, if any, in Q_A, Q_B, Q_C and Q_D. For counts 10–15, the states of Q_A, etc. are don't-cares and are shown by the symbol X. Using the J-K flip-flop transition table (Table 5.3) the required *J* and *K* conditions needed to cause each required transition can be filled in the third columns. For example, if the present count is $2 = 0010$, the next count is $3 = 0011$, and the required *J* and *K* inputs are

a) Q_A must change from 0 to 1, hence $J_A = 1$, $K_A = X$.

b) Q_B must remain at 1, hence $J_B = X$, $K_B = 0$.

c) Q_C and Q_D must remain at 0, hence $J_C = J_D = 0$ and $K_C = K_D = X$.

Note that, when the next state is a don't-care, all the *J* and *K* inputs are also don't-cares.

The next step is to draw a number of state diagrams, one for each *J* input and for each *K* input. These diagrams map Q_A, Q_B, Q_C and Q_D for each *J* or K value. The mapped equations are simplified by looping the appropriate squares with the results shown.

Map 1 (Q_A | $\bar{Q}_A$ columns; Q_C, $\bar{Q}_C$ rows)

Q_A		$\bar{Q}_A$		
X	X	X	X	Q_D
X	X	1	1	$\bar{Q}_D$
X	X	1	1	
X	X	1	X	Q_D
Q_B	$\bar{Q}_B$	Q_B		

1 $J_A = 1$

Map 2

Q_A		$\bar{Q}_A$		
X	X	X	X	Q_D
1	1	X	X	
1	1	X	X	$\bar{Q}_D$
X	1	X	X	Q_D
Q_B	$\bar{Q}_B$	Q_B		

2 $K_A = 1$

Map 3

Q_A		$\bar{Q}_A$		
X	X	X	X	Q_D
X	1	0	X	$\bar{Q}_D$
X	1	0	X	
X	0	0	X	Q_D
Q_B	$\bar{Q}_B$	Q_B		

3 $J_B = Q_A \bar{Q}_D$

Map 4

Q_A		$\bar{Q}_A$		
X	X	X	X	Q_D
1	X	X	0	
1	X	X	0	$\bar{Q}_D$
X	X	X	X	Q_D
Q_B	$\bar{Q}_B$	Q_B		

4 $K_B = Q_A$

Map 5

Q_A		$\bar{Q}_A$		
X	X	X	X	Q_D
X	X	X	X	$\bar{Q}_D$
1	0	0	0	
X	0	0	X	Q_D
Q_B	$\bar{Q}_B$	Q_B		

5 $J_C = Q_A Q_B$

Map 6

Q_A		$\bar{Q}_A$		
X	X	X	X	Q_D
1	0	0	0	$\bar{Q}_D$
X	X	X	X	
X	X	X	X	Q_D
Q_B	$\bar{Q}_B$	Q_B		

6 $K_C = Q_A Q_B$

Map 7

Q_A		$\bar{Q}_A$		
X	X	X	X	Q_D
1	0	0	0	$\bar{Q}_D$
0	0	0	0	
X	X	X	X	Q_D
Q_B	$\bar{Q}_B$	Q_B		

7 $J_D = Q_A Q_B Q_C$

Map 8

Q_A		$\bar{Q}_A$		
X	X	X	X	Q_D
X	X	X	X	$\bar{Q}_D$
X	X	X	X	
X	1	0	X	Q_D
Q_B	$\bar{Q}_B$	Q_B		

8 $K_D = Q_A$

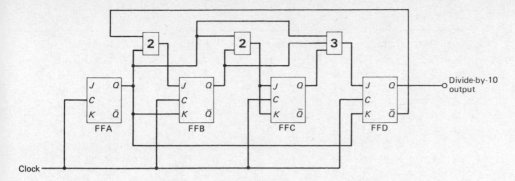

Fig. 5.15 Synchronous decade counter

The designed circuit is given in Fig. 5.15.

The design can be checked using a *state checking table* (Table 5.5).

1 The count starts at 0 so write $Q_A = Q_B = Q_C = Q_D = 0$.

2 Determine from the circuit the J and K values of each flip-flop at this time. Since $Q_A = 0$, then $J_B = K_B = J_C = K_B = J_D = K_D = 0$.

3 Now determine the next state. Since K_A and J_A are at 1, only the first stage will toggle so that the next state is 0001 or denary 1.

4 Write down the next state of the previous line as the present state of the next line and then find the values of each J and K input. Now $K_B = K_D = 1$ since $Q_A = 1$. Also, since $\bar{Q}_D = 1$, both inputs to the left-hand AND gate are 1 and thus $J_D = 1$. All other J and K inputs are at 0.

5 Determine the next state. Both flip-flops A and B toggle so that the next state is 0010 or denary 2.

6 This procedure is repeated, line-by-line, until a previous count is obtained. In this case this is 0000 which occurs on the tenth line.

As a further example consider the design of a synchronous counter to follow the sequence 0, 8, 4, 2, 9, 12, 14, 15, 7, 3, 1, 0, etc. The state table for the required counter is given by Table 5.6.

For the other numbers, i.e. 5, 6, 10, 11, and 13 the J and K inputs are all don't-cares.

Table 5.5 State checking table

	Present state												Next state			
	Q_D	Q_C	Q_B	Q_A	J_D	K_B	J_C	K_C	J_B	K_B	J_A	K_A	Q_D	Q_C	Q_B	Q_A
0	0	0	0	0	0	0	0	0	0	0	1	1	0	0	0	1	1
1	0	0	0	1	0	1	0	0	1	1	1	1	0	0	1	0	2
2	0	0	1	0	0	0	0	0	0	0	1	1	0	0	1	1	3
3	0	0	1	1	0	1	1	1	1	1	1	1	0	1	0	0	4
4	0	1	0	0	0	0	0	0	0	0	1	1	0	1	0	1	5
5	0	1	0	1	0	1	0	0	1	1	1	1	0	1	1	0	6
6	0	1	1	0	0	0	0	0	0	0	1	1	0	1	1	1	7
7	0	1	1	1	1	1	1	1	1	1	1	1	1	0	0	0	8
8	1	0	0	0	0	0	0	0	0	0	1	1	1	0	0	1	9
9	1	0	0	1	0	1	0	0	0	1	1	1	0	0	0	0	0

Table 5.6

	Present state					Next state					Required J-K inputs							
	Q_D	Q_C	Q_B	Q_A		Q_D^+	Q_C^+	Q_B^+	Q_A^+		J_D	K_D	J_C	K_C	J_B	K_B	J_A	K_A
0	0	0	0	0	8	1	0	0	0		1	X	0	X	0	X	0	X
8	1	0	0	0	4	0	1	0	0		X	1	1	X	0	X	0	X
4	0	1	0	0	2	0	0	1	0		0	X	X	1	1	X	0	X
2	0	0	1	0	9	1	0	0	1		1	X	0	X	X	1	1	X
9	1	0	0	1	12	1	1	0	0		X	0	1	X	0	X	X	1
12	1	1	0	0	14	1	1	1	0		X	0	X	0	1	X	0	X
14	1	1	1	0	15	1	1	1	1		X	0	X	0	X	0	1	X
15	1	1	1	1	7	0	1	1	1		X	1	X	0	X	0	X	0
7	0	1	1	1	3	0	0	1	1		0	X	X	1	X	0	X	0
3	0	0	1	1	1	0	0	0	1		0	X	0	X	X	1	X	0
1	0	0	0	1	0	0	0	0	0		0	X	0	X	0	X	X	1

From the table the state diagrams for each J and K input are as follows:

	Q_A		$\bar{Q}_A$		
Q_C	X	X	0	1	Q_D
	X	X	0	X	$\bar{Q}_D$
$\bar{Q}_C$	X	X	0	1	
	X	X	0	X	Q_D
	Q_B	$\bar{Q}_B$	Q_B		

1 $J_A = Q_B$

	Q_A		$\bar{Q}_A$		
Q_C	X	1	X	X	Q_D
	X	X	1	X	$\bar{Q}_D$
$\bar{Q}_C$	X	0	0	X	
	X	0	0	X	Q_D
	Q_B	$\bar{Q}_B$	Q_B		

2 $J_B = Q_C$

	Q_A		$\bar{Q}_A$		
Q_C	X	X	X	X	Q_D
	X	X	X	X	$\bar{Q}_D$
$\bar{Q}_C$	0	0	0	0	
	X	1	1	X	Q_D
	Q_B	$\bar{Q}_B$	Q_B		

3 $J_C = Q_D$

	Q_A		$\bar{Q}_A$		
Q_C	X	X	X	X	Q_D
	0	X	0	X	$\bar{Q}_D$
$\bar{Q}_C$	0	0	1	1	
	X	X	X	X	Q_D
	Q_B	$\bar{Q}_B$	Q_B		

4 $J_D = \bar{Q}_A \bar{Q}_C$

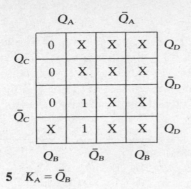

	Q_A			$\bar{Q}_A$	
Q_C	0	X	X	X	Q_D
	0	X	X	X	$\bar{Q}_D$
$\bar{Q}_C$	0	1	X	X	
	X	1	X	X	Q_D
	Q_B	$\bar{Q}_B$	Q_B		

5 $K_A = \bar{Q}_B$

	Q_A			$\bar{Q}_A$	
Q_C	0	X	X	0	Q_D
	0	X	X	X	$\bar{Q}_D$
$\bar{Q}_C$	1	X	X	1	
	X	X	X	X	Q_D
	Q_B	$\bar{Q}_B$	Q_B		

6 $K_B = \bar{Q}_C$

	Q_A			$\bar{Q}_A$	
Q_C	0	X	0	0	Q_D
	1	X	1	X	$\bar{Q}_D$
$\bar{Q}_C$	X	X	X	X	
	X	0	X	X	Q_D
	Q_B	$\bar{Q}_B$	Q_B		

7 $K_C = \bar{Q}_D$

	Q_A			$\bar{Q}_A$	
Q_C	1	X	0	0	Q_D
	X	X	X	X	$\bar{Q}_D$
$\bar{Q}_C$	X	X	X	X	
	X	0	1	X	Q_D
	Q_B	$\bar{Q}_B$	Q_B		

8 $K_D = Q_A Q_C + \bar{Q}_A \bar{Q}_C$

The required circuit is shown in Fig. 5.16

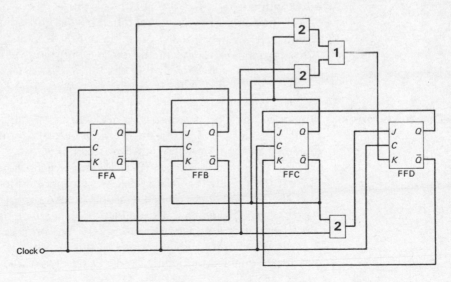

Fig. 5.16 Synchronous counter with a count of 0, 8, 4, 2, 9, 12, 14, 15, 7, 3, 1, 0, etc.

Use of D Flip-Flops

The design of a synchronous counter using **D flip-flops** is somewhat simpler than that described for a J-K flip-flop although the basic idea is the same. The simplification arises because the next-state outputs of each D flip-flop are *always* the same as the present-state D inputs. If the decade counter was to be re-designed with D flip-flops the first two lines in the state table would be

Present state				Next state							
Q_D	Q_C	Q_B	Q_A	D_D	D_C	D_B	D_A	Q_D^+	Q_B^+	Q_C^+	Q_A^+
0	0	0	0	0	0	0	0	1	0	0	0	1
1	0	0	0	1	0	0	1	0	0	0	1	0

If the design is proceeded with it will be found that more gates are needed than for the J-K version.

Synchronous Integrated Circuit Counters

The ttl family includes the following synchronous counters:
   74160 4-bit decade counter with direct clear.
   74161 4-bit binary counter with direct clear.
   74162 4-bit decade counter.
   74163 4-bit binary counter.
These devices are available in both the standard and the low-power Schottky types (including ALS).

All of the devices listed use T flip-flops. For the first two, a low level at the non-synchronous clear input will reset all the stages to have a low Q output whatever the states of the clock or the enable inputs. For the last two devices, the clear facility is synchronous, a low level at the clear input resets all Q outputs low *after* the next clock pulse. The synchronous clear facility allows the count to be easily modified because decoding the maximum count can then be achieved using only the one NAND gate.

Synchronous counters in the cmos family are of the *Johnson* type and these are discussed later in this chapter.

Up-Down Counters

For some digital applications it is necessary to be able to count downwards, i.e. 9, 8, 7, 6, 5, 4, 3, 2, 1, and 0. Many circuits are able to count in either direction and are known as **up-down counters**.

It was mentioned on page 103 that a non-synchronous counter would count down if the $\bar{Q}$ output of each stage were connected to the clock input of the next stage. The change-over from one function to the other is arranged by means of AND gates that are enabled by a *count-up* or a *count-down* pulse.

The circuit of a **non-synchronous** 3-bit **up-down counter** is shown in Fig. 5.17. If the count-up line is taken to the logical 1 level, the AND gates A and D are enabled and connect the Q outputs of the flip-flops A and B to

Fig. 5.17 Up-down counter

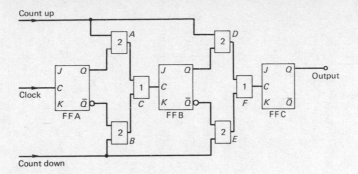

the clock input of the following flip-flops. The circuit then operates as an up-counter with a count of 8.

When only the count-down line is taken to logical 1, gates B and E are enabled, while gates A and D are inhibited. Now the $\bar{Q}$ outputs of the flip-flops A and B are connected to the clock inputs of the following stages and the circuit acts as a down-counter, again with a count of 8.

Table 5.7 *State table for divide-by-7 up-down counter*

Present state			Count-up line	Next state			Required J-K inputs					
Q_C	Q_B	Q_A	Z	Q_C^+	Q_B^+	Q_A^+	J_C	K_C	J_B	K_B	J_A	K_A
0	0	0	0	0	0	1	0	X	0	X	1	X
0	0	0	1	1	1	0	1	X	1	X	0	X
0	0	1	0	0	1	0	0	X	1	X	X	1
0	0	1	1	0	0	0	0	X	0	X	X	1
0	1	0	0	0	1	1	0	X	X	0	1	X
0	1	0	1	0	0	1	0	X	X	1	1	X
0	1	1	0	1	0	0	1	X	X	1	X	1
0	1	1	1	0	1	0	0	X	X	0	X	1
1	0	0	0	1	0	1	X	0	0	X	1	X
1	0	0	1	0	1	1	X	1	1	X	1	X
1	0	1	0	1	1	0	X	0	1	X	X	1
1	0	1	1	1	0	0	X	0	0	X	X	1
1	1	0	0	0	0	0	X	1	X	1	0	X
1	1	0	1	1	0	1	X	0	X	1	1	X

The design of a **synchronous up-down counter** follows similar lines to that for an up-counter. Suppose, for example, that a divide-by-7 up-down counter is wanted. Assume that, when the count-up line is at 1, the count-down line will be at 0 and vice versa.

The state table of the required counter is given in Table 5.7. From the state table the state diagrams for each of the six J and K inputs can be written down (Z denotes the state of the count-up line):

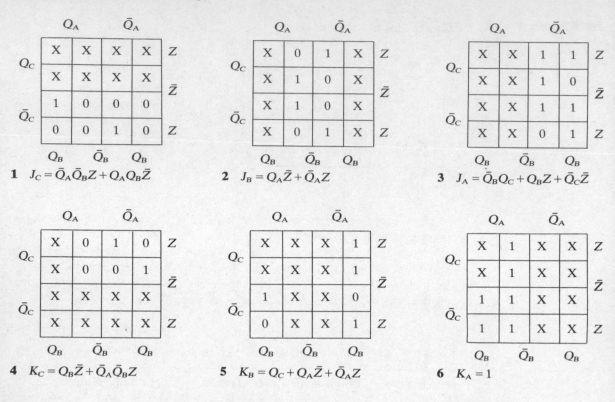

1 $J_C = \bar{Q}_A\bar{Q}_BZ + Q_AQ_B\bar{Z}$

2 $J_B = Q_A\bar{Z} + \bar{Q}_AZ$

3 $J_A = \bar{Q}_BQ_C + Q_BZ + \bar{Q}_C\bar{Z}$

4 $K_C = Q_B\bar{Z} + \bar{Q}_A\bar{Q}_BZ$

5 $K_B = Q_C + Q_A\bar{Z} + \bar{Q}_AZ$

6 $K_A = 1$

Fig. 5.18

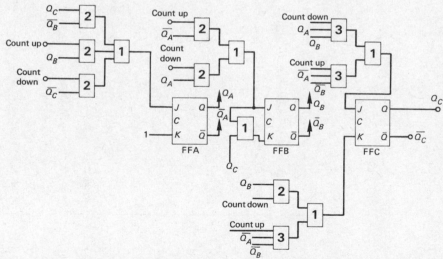

The required circuit is shown in Fig. 5.18.

Synchronous up-down counters available in the ttl family are the 74190/1/2/3. These are, respectively, bcd, binary, decade, and 4-bit binary counters and they are manufactured in both the standard and the low-power Schottky versions. The first two circuits incorporate J-K flip-flops but the other two use T flip-flops.

Shift Registers

A **shift register** consists of a number of J-K flip-flops connected in cascade as shown by Fig. 5.19a. The connection of Q_A to J_B and of $\bar{Q}_A$ to K_B and so on ensures that each flip-flop will take up the state, $Q = 0$ *or* $Q = 1$, of the preceding flip-flop at the end of each clock pulse. Each flip-flop transfers its *bit* of information to the following flip-flop whenever a clock pulse occurs. Alternatively, D flip-flops can be used with the Q output of each stage connected to the D input of the next stage (see Fig. 5.19b). The flip-flops are usually provided with a clear or reset terminal so that the register can be cleared or set to 0. Shift registers in the cmos family generally use D flip-flops, whereas the ttl versions usually employ S-R flip-flops.

Suppose that, initially, all four flip-flops are cleared, i.e. $Q_A = Q_B = Q_C = Q_D = 0$. If the data to be stored by the register is applied serially to the input terminal, the input data will shift one stage to the right each time a clock pulse arrives. This means that, if the data stored in the register is a binary number, the most significant bit is on the right. If the data applied to the input terminal is 11010, the action of the circuit is as follows.

At the end of the first clock pulse, $Q_A = 1$. After the second pulse, $Q_A = 0$, $Q_B = 1$. The third clock pulse will cause $Q_C = Q_A = 1$ but the second flip-flop will now reset to have $Q_B = 0$. The fourth bit of data is a 1 and this will be stored by flip-flop A at the end of the fourth clock pulse, when the bits stored by the other flip-flops all move one place to the right. At this stage the data stored is 1010. There is one more bit, a 1, to be applied to the input terminal of the register. At the end of the next clock pulse, flip-flop A sets and all the other flip-flops take up the state of the preceding stage. Now the complete number is stored by the register.

Fig. 5.19 (a) Shift register using J-K flip-flops, (b) shift register using D flip-flops

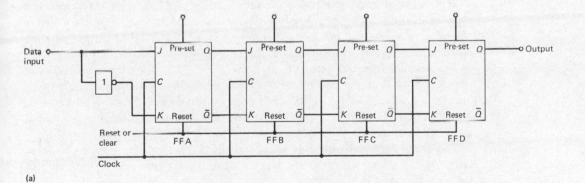

(a)

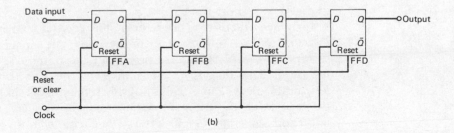

(b)

Fig. 5.20 Timing diagram for a 4-bit shift register

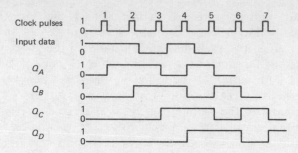

The timing diagram illustrating the operation of the register is shown in Fig. 5.20. Note that the Q_A waveform is delayed behind the input data waveform by a time that may not be equal (in the figure it isn't) to the periodic time of the clock waveform. However, the Q waveforms of all the other flip-flops are delayed behind the Q waveform of the preceding flip-flop by *exactly* the clock period.

The shift register can be used to multiply or divide a binary number by a factor of 2 by merely shifting the number one place to the right or to the left, respectively. In either case any part of the number must not be shifted right out of the register.

In general, a shift register can be used in any one of four different ways. These methods are known as

1 Serial-in Parallel-out (SIPO)
2 Parallel-in Serial-out (PISO)
3 Serial-in Serial-out (SISO)
4 Parallel-in Parallel-out (PIPO)

With a **serial-in/parallel-out** shift register (Fig. 5.21a), data is fed into the register in the manner previously described and, when the complete word is stored, all the bits can be read off simultaneously from the Q output of each stage. The register acts to convert data from serial into parallel form.

The **parallel-in/serial-out** shift register (Fig. 5.21b) operates in exactly the opposite way. The data to be stored is set up by first clearing all the stages and then applying logical 1 to the preset terminal of each stage which is to be set. The data can then be shifted out of the register, one bit at a time, under the control of the clock.

The **serial-in/serial-out** shift register (Fig. 5.21c) can be used as a delay circuit, or as a short-term store, but the stored data can only be accessed in the order in which it is stored.

Fig. 5.21d shows a **parallel-in/parallel-out** shift register and this circuit also acts as a short-term store.

Sometimes there is a need for a shift register that has the capability to move data *either* to the left *or* to the right. The circuit of such a register is shown in Fig. 5.22. There are two data input terminals, one of which is used for serial data that is to be shifted to the right and the other is for left-shifting data. The direction in which the data is shifted is determined by the logic levels on the two lines marked, respectively, as Shift right 1/shift left 0 and Shift right 0/shift left 1. The circuit is arranged so that the signals

Fig. 5.21 Methods of
using shift registers
  (*a*) Serial-in/
       parallel-out
  (*b*) Parallel-in/
       serial-out
  (*c*) Serial-in/
       serial-out
  (*d*) Parallel-in/
       parallel-out

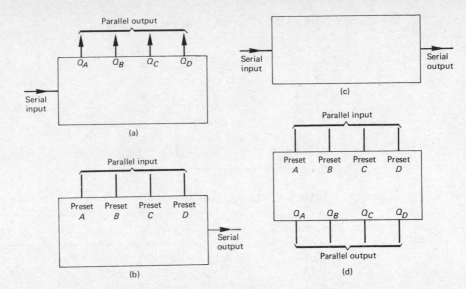

(a)

(b)

(c)

(d)

Fig. 5.22 Shift-
right/shift-left register

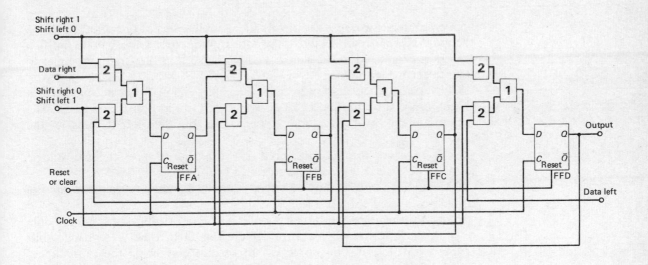

Table 5.8 TTL shift registers

Number	Type	Shift
7491	8-bit SISO	Right
7496	5-bit PISO	Right
74164	8-bit SIPO	Right
74165	4-bit PISO	Right
74194	4-bit PIPO	Left and right
74195	4-bit PIPO	Right
74198	8-bit Universal	Left and right

Table 5.9 CMOS shift registers

Number	Type	Shift
4006	18-stage static SISO	Right
4014	8-stage static PISO or SISO	Right
4015	dual 4-bit static SIPO	Right
4021	8-stage static PISO or SISO	Right
4031	64-stage static SISO	Right
4035	4-stage static PIPO	Right
4062	200-stage dynamic SISO	Right
40100	32-stage static SISO	Left and right
40194	4-stage static Universal	Left and right

applied to these two lines are always the complements of one another. When the top AND gates are enabled, the data-right input is connected to flip-flop A. Q_A is connected to flip-flop B and so on, so that the circuit is similar to that given in Fig. 5.19b.

Conversely, when the lower AND gates are enabled, Q_D is connected to the D terminal of flip-flop C, Q_C is connected to flip-flop B and so on. The circuit will then shift data entered serially at the data-left terminal to the left.

Shift registers are manufactured in both the ttl and the cmos logic families. Some examples in the ttl range are given in Table 5.8.

CMOS shift registers may be either static or dynamic types and examples of devices are given in Table 5.9.

Static shift registers store their data in a number of flip-flops and are able to hold the information for an indefinite length of time. A dynamic shift register, on the other hand, stores bits of information on the capacitances linking the stages of the register. Because the charge held in these capacitances gradually leaks away, the dynamic shift register must be periodically *refreshed*. The dynamic type of shift register therefore possesses the disadvantages of needing more complex supporting circuitry and a minimum clock frequency, but it does have the important advantage of a very much higher storage capability. This is made evident by Table 5.9 from which it can be seen that the 4062 dynamic shift register has 200 stages, compared with up to 64 stages for the static type. Other dynamic shift registers may have even larger storage capacities, for example, 512 stages or 1024 stages. Clearly, all large capacity shift registers must be of the serial-out type because the number of ic package pins is limited.

Dynamic CMOS Shift Registers

Shift registers constructed by the interconnection of flip-flops as previously described are not practical when high bit storage capabilities are required. This is because the area of chip occupied would be too large and the power dissipation would be excessively high. Large capacity shift registers— perhaps hundreds of bits—are therefore always of the dynamic type.

Essentially, a dynamic shift register stage is fabricated by the cascade of two dynamic mos inverters, bit storage being *temporarily* achieved by charging the gate-substrate capacitance of a mosfet.

Fig. 5.23 CMOS dynamic shift register stage

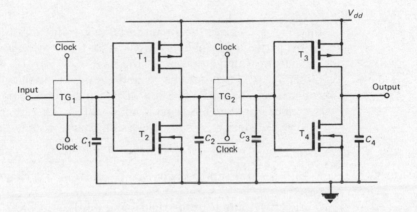

The circuit of a **dynamic cmos shift register** is shown in Fig. 5.23. The boxes marked as TG represent *transmission gates* (p. 52) that are controlled by complementary clock signals. Transistors T_1 and T_2, and transistors T_3 and T_4, form two cmos inverter stages.

When the clock is at the logical 1 voltage level, TG_1 is turned ON and the input signal is applied to T_1/T_2 *and* across the capacitance C_1. When the clock becomes 0, TG_1 turns OFF and TG_2 turns ON; capacitances C_2 and C_3 are now effectively placed in parallel and the voltage developed across them is inverted by T_3/T_4 to appear across the output capacitance C_4. This means that the input signal appears at the output of the circuit after a time delay equal to the periodic time of the clock.

Although the cmos shift register contains eight mosfets in all, the total power dissipation is very small since the only currents that flow are the minute ones needed to charge the four capacitances. These capacitances are only of the order of 0.5 pF.

NMOS dynamic shift registers are also used when very large storage capacities are needed and these may well be able to store more than 1000

Fig. 5.24 NMOS shift
register stage

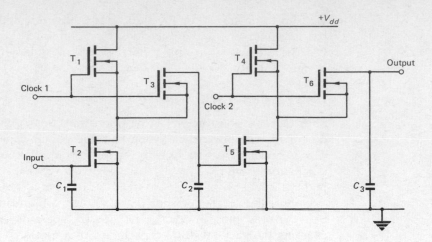

bits. Fig. 5.24 gives the circuit of an nmos shift register stage; T_1 and T_4 act
as the active loads. When a clock is at 1, the associated mosfets are turned
ON and the voltage present at the gate of T_2, or T_5, will be inverted. Also,
either T_3 or T_6 will also be ON and the inverted voltage will charge up either
capacitance C_2 or C_3. The rest of the circuit operation is very similar to that
described for the cmos circuit and is left as an exercise (Exercise 5.26).

Shift registers are also made using charge-coupled devices.

Ring Counters

A shift register can be connected to function as a **ring counter** by connecting
together

a) the Q output of the right-hand flip-flop to the J input of the left-hand
flip-flop and

b) the $\bar{Q}$ output of the right-hand flip-flop to the K input of the left-hand
flip-flop.

The circuit of a ring counter is shown in Fig. 5.25.

If D flip-flops were used, the Q output of flip-flop D would be connected
to the D input of flip-flop A. The internal states of the register will now be
circulated around the loop and, for an *n-stage register*, the binary pattern
will be repeated at the output after every n clock pulses. The ring counter
can therefore be used as a *delay line*; the data is stored and circulated
internally in a multiple bit pattern and any particular bits can be read out as
and when required.

If only a single bit is circulated, the output of each flip-flop will give a
uniquely timed 1-*out-of-n count*. The count is read by finding the flip-flop
that is set; no decoding is needed. Consider, for example, a 5-stage register;
the pattern held is given in Table 5.10 and the timing diagram is shown by
Fig. 5.26*a*. If the flip-flops A *and* B are initially set (Fig. 5.26*b*), a double-
width pulse will circulate around the register.

The basic ring counter is not self-starting but it can easily be modified to

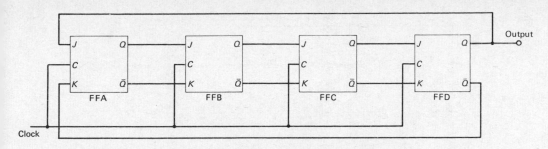

Fig. 5.25 Ring counter

Fig. 5.26 Timing diagram for a 5-stage shift register

Table 5.10   5-stage register state table

Clock pulse	Q_E	Q_D	Q_C	Q_B	Q_A
	0	0	0	0	1
1	0	0	0	1	0
2	0	0	1	0	0
3	0	1	0	0	0
4	1	0	0	0	0
5	0	0	0	0	1

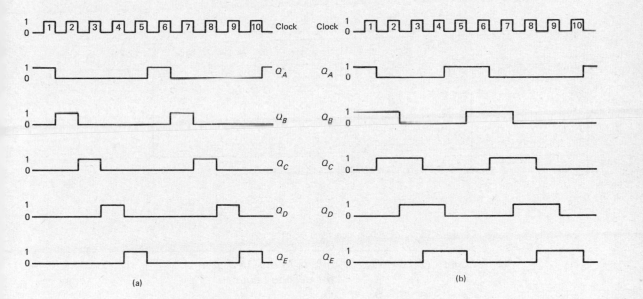

(a) (b)

become so by the addition of a NOR gate and an inverter (which could, of course, be another NOR gate). The method is illustrated by Fig. 5.27. The circuit need not be set initially. No matter what the initial state happens to be, the counter will go to the "all 0" state *before* going into the repetitive pattern, because only then will the feedback be 1. A 10-stage ring counter can be used as a decade counter that requires no feedback logic or decoding but for some applications its lack of a binary output may be a disadvantage.

Fig. 5.27 Self-starting
ring counter

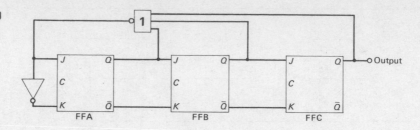

Johnson Counter

A **Johnson counter** or **twisted ring counter** differs from the ring counter in
that the Q and $\bar{Q}$ connections of the final flip-flop are interchanged, or
twisted (see Fig 5.28). The sequence of the Johnson counter is given by
Table 5.11. Decoding of each count is easy since it only needs one 2-input
AND gate for each count.

Fig. 5.28 Johnson
counter

The Johnson counter possesses the advantages of (i) high speed of
operation because it does not follow the true 8421 binary code and (ii) the
decoded outputs are glitch-free because any output is subject to only one

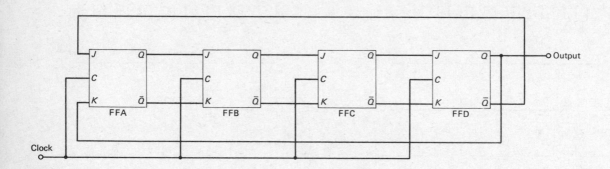

Table 5.11 Johnson counter sequence

Count	Q_D	Q_C	Q_B	Q_A	Inputs to the AND gate for decoded count
0	0	0	0	0	$\bar{Q}_A\bar{Q}_C$
1	0	0	0	1	$Q_A\bar{Q}_B$
2	0	0	1	1	$Q_B\bar{Q}_C$
3	0	1	1	1	$Q_C\bar{Q}_D$
4	1	1	1	1	Q_AQ_C
5	1	1	1	0	$\bar{Q}_AQ_B$
6	1	1	0	0	$\bar{Q}_BQ_C$
7	1	0	0	0	$\bar{Q}_CQ_D$
8	0	0	0	0	$\bar{Q}_A\bar{Q}_C$

Fig. 5.29 Possible states for a 4-stage Johnson counter

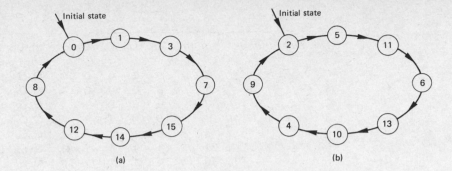

stage's delay. The disadvantage of the circuit is that there exists the possibility of the counter locking into an unwanted state.

If all the stages are initially in either the 0 or the 1 state, the number of possible states is $2n$, where n is the number of stages. Thus, referring again to the 4-stage counter, the 8 possible states are shown in Fig. 5.29a when the initial count is 0 (as in Table 5.11) and in Fig. 5.29b when the initial count is 2.

Similarly, the four groups of possible states for a 5-stage Johnson counter are given in Fig. 5.30. Notice that the *total* number of different states is 32 or 2^5. For the first cycle (Fig. 5.30a), the required output decoding is given in Table 5.12.

Fig. 5.30 4 cycles of possible states for a 5-stage Johnson counter

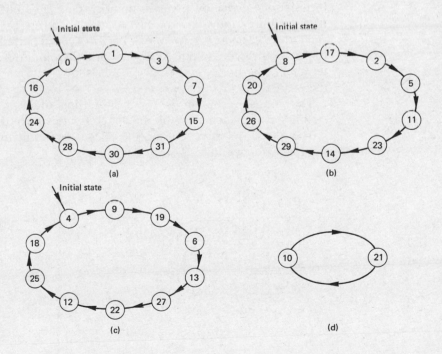

Table 5.12

Decimal number	Count	Q_E	Q_D	Q_C	Q_B	Q_A	Inputs to AND gate for decoded count
0	0	0	0	0	0	0	$\bar{Q}_A\bar{Q}_E$
1	1	0	0	0	0	1	$Q_A\bar{Q}_B$
2	3	0	0	0	1	1	$Q_B\bar{Q}_C$
3	7	0	0	1	1	1	$Q_C\bar{Q}_D$
4	15	0	1	1	1	1	$Q_D\bar{Q}_E$
5	31	1	1	1	1	1	$Q_A Q_E$
6	30	1	1	1	1	0	$\bar{Q}_A Q_B$
7	28	1	1	1	0	0	$\bar{Q}_B Q_C$
8	24	1	1	0	0	0	$\bar{Q}_C Q_D$
9	16	1	0	0	0	0	$\bar{Q}_D Q_E$

The cmos 4017 is a 5-stage Johnson decade counter that has ten decoded outputs (Fig. 5.31a), each of which can drive other circuits. If one of the outputs is connected to the clock inhibit terminal, the count can be stopped at any desired point. The outputs go high in sequence, one at a time, under the control of the clock. The terminal marked as Carry out goes low after five, and then after ten, clock pulses and can be connected to the clock terminal of another 4017 if a count greater than 10 is wanted. The reset terminal can be used to set the 0 output to 1 and hence all the other outputs to 0. The timing diagram of a 4017 is shown in Fig. 5.31b.

Another useful presettable cmos Johnson counter is the 4018 (Fig. 5.31c). This is a 5-stage circuit that can be used to divide by any ratio from 2 to 10 inclusive. To divide by any *even* number, the $\bar{Q}_n$ output, where n is *one-half* of the required ratio, is connected to the data terminal. To divide by any *odd* ratio, the $\bar{Q}_m$ and $\bar{Q}_n$ terminals, where $m + n = $ required ratio, are connected to the inputs of a 2-input AND gate and the output of this gate to the data terminal. The preset and enable signal allows data on the J inputs to reset the counter.

The Johnson counter is not self-starting unless some extra feedback circuitry is provided. Suitable feedback for this purpose can be designed.

The Karnaugh map showing the decimal equivalents of the states in the wanted sequence for a 4-stage counter is

The squares representing unused states have been marked U. The

Fig. 5.31 (*a*) Pin connections of the 4017 5-stage Johnson counter
   (*b*) 4017 timing diagram
   (*c*) Pin connections of the 4018 presettable 5-stage Johnson counter

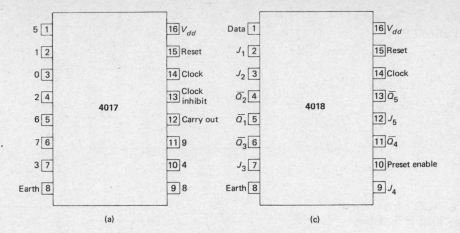

(a)   (c)

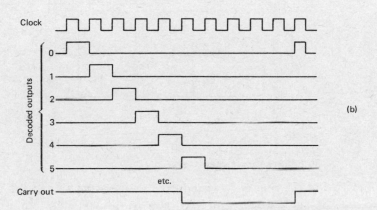

(b)

Boolean expression that represents these states is

$$F = Q_A Q_C \bar{Q}_B + \bar{Q}_A Q_C \bar{Q}_D + Q_A \bar{Q}_C Q_D + \bar{Q}_A Q_B \bar{Q}_C$$

The logic of these states can be used to clear the counter if a wrong sequence is entered.

To make the counter self-starting, logic representing one state from each possible unwanted sequence should be used. In the case of the 4-stage Johnson counter there is only the one other possible sequence. Thus, if, for example, decimal number 5 is chosen, the required logic is $F = Q_A \bar{Q}_B Q_C \bar{Q}_D$. If the output of the AND gate with this input is used to clear the counter, the circuit will self-start within 8 clock pulses.

Fig. 5.32 Feedback shift register

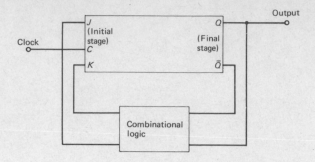

Fig. 5.33 Linear feedback shift register

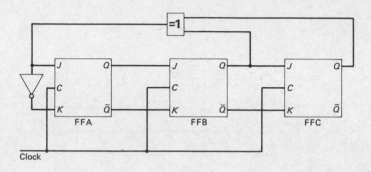

Feedback Shift Registers

In the ring and twisted ring counters the Q and the $\bar{Q}$ outputs of the final stage are taken directly back to the first stage. In a **feedback shift register** the feedback is passed through a combinational logic circuit (see Fig. 5.32). If the combinational logic circuit consists *only* of exclusive-OR gates, the circuit is said to be a *linear feedback shift register*.

Suppose, for example, that a three-stage circuit and one exclusive-OR gate are used with inputs Q_B and Q_C (Fig. 5.33). The output of the gate, which is applied to the J terminal of the first stage, is

$$J_A = Q_B \bar{Q}_C + \bar{Q}_B Q_C$$

The initial set is $Q_A = 1$, $Q_B = Q_C = 0$, hence $J_A = 0$, $J_B = 1$. At the end of the first clock pulse, Q_A becomes 0 and Q_B becomes 1, Q_C remains unchanged at 0. Now $J_A = 1.1 + 0.0 = 1$, $J_B = 0$, and $J_C = 1$. At the end of the next clock pulse, Q_A becomes 1 again, $Q_B = 0$, and $Q_C = 1$. Now $J_A = 0.0 + 1.1 = 1$, $J_B = 1$, and $J_C = 0$, and so the next state is $Q_A = Q_B = 1$ and $Q_C = 0$. The operation is summarized by Table 5.13.

The "all Qs zero" state must be avoided because it would lock the J_A input at 0. This requirement can be achieved by ANDing all the $\bar{Q}$ outputs and applying the output of the AND gate to the J_A input. This means that the maximum cycle length is $2^n - 1$. The stages from which the feedback must be taken to give a maximum cycle length are shown in Table 5.14.

Table 5.13 State table of linear feedback register

Q_C	Q_B	Q_A
0	0	1
0	1	0
1	0	1
0	1	1
1	1	1
1	1	0
1	0	0
0	0	1

Table 5.14

Number of stages	Linear feedback	Cycle length
2	$1 \oplus 2$	3
3	$1 \oplus 3$ or $2 \oplus 3$	7
4	$3 \oplus 4$ or $1 \oplus 4$	15
5	$3 \oplus 5$ or $2 \oplus 5$	31
6	$5 \oplus 6$	63
7	$4 \oplus 7$	127
8	$4 \oplus 5 \oplus 6 \oplus 8$	255
9	$5 \oplus 9$	511
10	$7 \oplus 10$	1023

(Note: $\oplus$ indicates exclusive-OR function.)

Table 5.15 State table of non-linear feedback register

State	Q_D	Q_C	Q_B	Q_A	Q_D^+	Q_C^+	Q_B^+	Q_A^+
0	0	0	0	0	0	0	0	1
1	0	0	0	1	0	0	1	0
2	0	0	1	0	0	1	0	1
5	0	1	0	1	1	0	1	0
10	1	0	1	0	0	1	0	0
4	0	1	0	0	1	0	0	0
8	1	0	0	0	0	0	0	0

Other feedback combinations will produce shorter cycle lengths.

The output of a linear fsr forms a pseudo-random binary sequence which, if there are a large enough number of stages, may approximate to the white noise power density spectrum. For example, from Table 5.13 it can be seen that the output of a 3-stage circuit is

100110111011101010001

This sequence contains all the possible 3-bit combinations *except* 000.

Fig. 5.34 Determination of the possible states in a non-linear feedback shift register

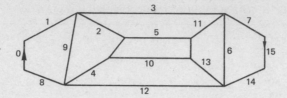

Fig. 5.35 Non-linear feedback shift register

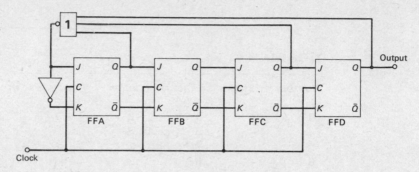

Non-linear Feedback Shift Registers

If the feedback from the Q terminals of a shift register to its input terminal does *not* consist exclusively of exclusive-OR gates, the circuit is said to be non-linear.

The two possible states of a 4-stage Johnson counter are shown in Figs. 5.29a and b. It is possible to move out of one cycle and into the other and back again at certain points. Fig. 5.34 shows the two cycles combined, with connections linking 2 and 4, and 10 and 11. There are 16 different states and so any count up to this value can be designed.

Suppose, for example, that a count of 7 using the sequence $0, 1, 2, 5, 10, 4, 8, 0 \ldots$ is required. The state table is given by Table 5.15. From the table

$$J_A = \bar{Q}_A \bar{Q}_B \bar{Q}_C \bar{Q}_D + \bar{Q}_A Q_B \bar{Q}_C \bar{Q}_D = \bar{Q}_A \bar{Q}_C \bar{Q}_D$$

The required circuit is shown in Fig. 5.35.

Exercises 5

5.1 Two J-K master-slave flip-flops are connected to form a modulo 3 counter. Draw the circuit and, with the aid of a truth table and a timing diagram, explain its operation.

5.2 Show how 5 J-K flip-flops could be connected to form a twisted ring counter. What would be the count of this circuit? With the aid of a timing diagram explain its action.

5.3 Draw the diagram, and explain the action, of a shift register. Explain its operation when storing the binary number 0010.

5.4 Design a ripple counter whose outputs follow the 8421 numbering system.

5.5 Draw diagrams to show how a 4-bit shift register can be converted into a) a 4-bit ring counter, b) an 8-bit ring counter.

5.6 Design a divide-by-5 up-counter using a state table and map.

5.7 Draw the circuit of, and explain the operation of, a dynamic shift register. State some factors that affect the clock frequency.

5.8 Design a synchronous counter to pass through the following states in the order given: 011, 100, 101, and 111.

5.9 A linear feedback shift register has 4 stages A, B, C and D. Feedback paths exist between the outputs of stages A and D to the input of stage A. Determine the code cycle that is generated.

5.10 Design a counter with three stages to cycle through the states $0, 2, 1, 7, 5, 6, 0, \ldots$.

5.11 Draw the block diagram of a 3-stage shift register. Show the following inputs: serial data in, clock and reset.

5.12 Design a non-linear fsr to count through 11 states.

5.13 Design a 5-stage Johnson decade counter with decoding for alternate states starting with $Q_A = Q_B = Q_C = Q_D = Q_E = 0$.

5.14 Design a synchronous counter which counts up from 0 to 14 inclusive and resets to 0.

5.15 Design a synchronous counter to count down from 9 to 0 then resets to 9.

5.16 Design a 4-bit counter to follow the sequence

$$3, 11, 0, 10, 9, 8, 4, 14, 15, 1, 7, 13, 5, 12, 6, 2, 3 \ldots$$

5.17 Describe the operation of the circuit given in Fig. 5.24.

Short Exercises

5.18 Draw a diagram to show how a decade counter may be made using four flip-flops.

5.19 A non-synchronous counter is to count up to binary 1111 and then reset. Draw the block diagram of the circuit stating which type of flip-flops you are using.

5.20 Explain the advantages of synchronous counters over non-synchronous counters.

5.21 Show how a) a D and b) a J-K flip-flop can be connected as a divide-by-2 circuit. Explain the operation of the circuit with the aid of timing diagrams.

5.22 Draw the diagram of a ripple counter with bcd outputs.

5.23 Determine the count sequence of the divide-by-9 circuit of Fig. 5.14a. Note that it does not go from 0 to 8.

5.24 Show how two 7490 decade counters can be connected to give a count of 25.

5.25 Draw the diagram of a divide-by-200 circuit that employs two 7490 decade counters and one 7490 divide-by-12 counter.

5.26 Draw the circuit of, and explain the operation of, an nmos dynamic shift register.

5.27 How many stages are needed for a decade ring counter? Write down a table showing the counting sequence of such a circuit.

5.28 Draw circuits to show how two 4018 Johnson counters could be connected to give a division ratio of (i) 14 and (ii) 36.

6 Photo-electric Devices

Introduction

The action of a number of semiconductor devices is based upon the generation of hole-electron pairs when *light energy* is incident upon the semiconductor material. The photo-electric effects can be classified into three categories:

1   Photo-conductivity   The conductivity, or resistance, of the semiconductor material is altered by the incident light energy.

2   Photo-voltaic   An emf is generated whose magnitude is proportional to the intensity of the incident light.

3   Photo-emissivity   A current passing through the semiconductor material causes it to radiate energy in the visible section of the frequency spectrum.

Fig. 6.1, Spectral response of a photo-electric device

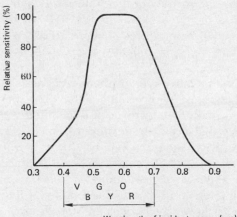

Any photo-electric device is sensitive to incident light, or it radiates light, only in a certain range of wavelengths. Its sensitivity is normally expressed in the form of a graph of output plotted to a base of wavelength. Such a graph is often known as the *spectral response* of the device and Fig. 6.1 gives a typical example. It can be seen that the device in question, actually a photo-diode, has its maximum sensitivity at wavelengths of about $0.53\ \mu$m to $0.66\ \mu$m or to green, yellow, orange and red light. In the case of a light-emitting device the graph would generally be referred to as its *emission spectrum* graph.

Light energy is measured in terms of units known as the *lumen* (lm), the candela (cd), the lux (lx), and the candela/m (cd/ms).

133

The total amount of light emitted, or received, by a surface is known as its *luminous flux* and its strength is measured in lumens. One lumen is the luminous flux emitted by a source of 1 candela within a solid angle of 1 steradian. A steradian is the solid angle subtended by the centre of a sphere of radius r by an area r^2 on the surface of the sphere. A complete sphere contains 4π steradians.

Luminous intensity is the strength of a light source in a given direction and it is measured in candela. Thus

Candela = lumens/steradian (6.1)

The luminous intensity of a light source may often vary with the angle from which the source is viewed. When the viewing angle is large, the term *luminance* is generally employed. Luminance is measured in cd/m and it is a measure of the brightness of a sphere.

The amount of luminous flux incident upon a surface is measured in lux (lx). 1 lx is the illumination produced when 1 lumen of light flux is incident upon a surface of area 1 m^2. An illumination of 1 lx is produced on an area of 1 m^2 at a distance of 1 m by a light source of 1 cd.

Photo-conductivity

The conductivity of a semiconductor material depends upon the concentration of the charge carriers in the material. If light energy is supplied to the material, more of its covalent bonds [EII] will be broken and extra hole-electron pairs will be produced. There is then an increase in the number of charge carriers and consequently the conductivity of the material is increased.

The **photo-conductive cell**, or **photo-resistor**, is a device that makes use of the phenomenon of photo-conductivity. The most commonly employed material for a photo-conductive cell is cadmium sulphide since this material gives a device that has (i) a maximum power dissipation of 300 mW or so, (ii) good sensitivity, and (iii) a high conductivity when there is incident light.

The symbol for a photo-conductive cell is given in Fig. 6.2*a*, while Fig. 6.2*b* shows a typical construction for such a device. Cadmium sulphide powder is sintered to produce a hard material and it is then formed into a zig-zag strip as shown. Electrodes are deposited at each end of the strip and then brought out to pins protruding through the base of the case.

The performance of a photo-conductive cell is often described in terms of such quantities as the *illuminated* and the *dark* resistances and the illumination sensitivity. The first two terms are, respectively, the resistances of the cell when it is illuminated by light and when it is in the dark. The last term is

$$\text{Illumination sensitivity} = \frac{\text{Illumination current (A)}}{\text{Incident lumination (lux)}} \qquad (6.2)$$

Fig. 6.3 shows how the resistance of a typical photo-conductive cell varies with the illumination; clearly cadmium sulphide is very nearly an insulating material when it is in the dark.

Fig. 6.2 (*a*) Symbol for a photo-conductive cell, (*b*) Typical construction

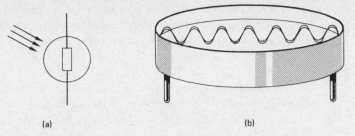

(a)

(b)

Fig. 6.3 Resistance of a photo-conductive cell

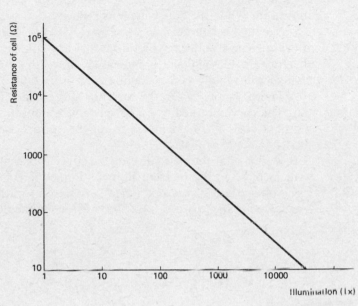

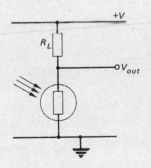

Fig. 6.4 Light measurement

Cadmium sulphide photo-conductive cells are used for a variety of purposes such as flame control or smoke detection units and as industrial on/off switches.

Some other materials are also used in photo-conductive cells designed for special applications. Indium antimonide is sensitive in the infra-red region—up to about 7.5 μm wavelength. Lead sulphide cells are usable at wavelengths of some 0.3–4 μm, while cadmium selenide is used as the cell material when the utmost sensitivity is required. This is particularly the case towards the blue/violet end of the visible spectrum. Cadmium selenide is also the choice if a rapid response is needed; for cadmium selenide the response time is about 10 ms as opposed to about 100 ms for cadmium sulphide. Unfortunately, the resistance of a cadmium selenide cell is temperature-dependent, while that of a cadmium sulphide cell is reasonably independent of temperature.

The photo-conductive effect can be used to measure the intensity of incident light using the circuit shown in Fig. 6.4. The output voltage of the circuit is given by

$$V_{out} = VR_{cell}/(R_L + R_{cell})$$

where R_{cell} is the resistance of the cell. The output voltage will vary as the value of R_{cell} is changed by the incident light. The output voltage can then be measured or it can be used to turn a transistor ON and OFF.

Photo-diodes

When a reverse-biased p-n junction is illuminated by visible light, the current that flows across the junction will vary, in a way that is very nearly linear, with the illumination. If the illumination is kept at a constant value it is found that the current increases only slightly as the reverse bias voltage is increased. When the bias voltage is reduced to 0 V and then has its polarity reversed, so that the p-n junction is forward biased, the current that flows will *not* change noticeably until the forward voltage reaches about 0.3 V. If the forward bias voltage is increased above this figure, the diode reverse current will rapidly fall, becoming zero at a voltage of about 0.5 V.

A typical family of reverse-current/voltage characteristics is shown in Fig. 6.5. The curve marked as "dark current" represents the current that flows when there is *no* incident light, i.e. the reverse saturation current of the diode.

Fig. 6.5 (*a*) Symbol for a photo-diode, (*b*) typical photo-diode characteristics

It is clear that the resistance of the reverse-biased photo-diode decreases with increase in the incident illumination. When, for example, the reverse bias voltage is 8 V, the resistance varies from $8/2.5 \times 10^{-6} = 3.2\,M\Omega$ when the diode is dark to $8/205 \times 10^{-6} = 39\,k\Omega$ when the illumination is 2000 lux.

(a)

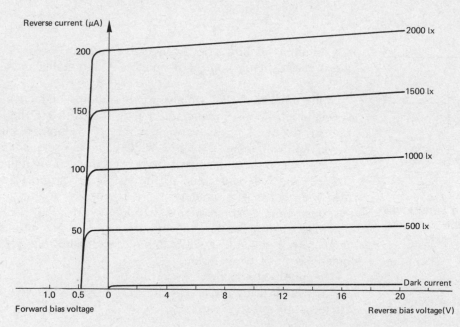

(b)

Example 6.1

A photo-diode with the I/V characteristics shown in Fig. 6.5 is connected in the circuit of Fig. 6.6b. The reverse bias voltage is 4 V and R is 10 000 Ω. Determine the voltage across the load resistor when the incident illumination is 1000 lx.

Solution
A dc load line should be drawn on the characteristics between the points $I = 0$, $V = 4$ V and $V = 0$, $I = 4/10\,000 = 400\ \mu A$. The load line cuts the curve for 1000 lx where the diode current is 100 μA and the diode voltage is 3.2 V.

Therefore, load voltage $= 100 \times 10^{-6} \times 10^{4} = 1$ V   (*Ans*)

Fig. 6.6 Photo-diode circuits

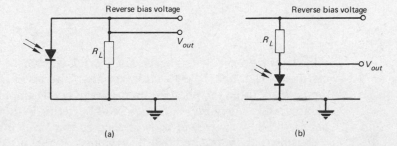

(a)   (b)

A photo-diode has two terminals, known respectively as the anode and the cathode, and it can be operated in its reverse-biased or its forward biased condition or sometimes with zero bias voltage. When the diode is reverse biased, it functions as a photo-conductive device; when it is forward biased it functions as a photo-voltaic device.

When a photo-diode is reverse biased, the photo-current generated by the incident light energy will develop a voltage across a load resistor connected in parallel with the diode (see Fig. 6.6a). A better circuit, since it results in less noise at the output, is shown in Fig. 6.6b. Either circuit can be used to measure light intensity or to operate a relay when the incident light intensity reaches some pre-determined value.

The manufacturers of photo-diodes quote various parameters for their devices and these include:

a) Sensitivity in nA/lx; generally quoted for a reverse-bias voltage some-what less than the maximum permitted value. A typical value is 11 nA/lx at $V_R = 15$ V for $V_{R(max)} = 18$ V.

b) The maximum permitted reverse bias voltage and reverse current; typically these are 20 V and 14 mA respectively.

c) The maximum forward current; typically 10 mA.

d) The spectral response; when plotted this is very similar to Fig. 6.1.

When a photo-diode is operated with a small forward-bias voltage, the reverse current falls rapidly and at some value, in the region of 0.3–0.5 V, becomes equal to zero. The actual voltage at which zero diode current flows is known as the *photo-voltaic voltage*. This is the voltage developed by the photo-diode when its terminals are open-circuited. Although not evident from Fig. 6.5 the photo-generated voltage increases slightly with increase in

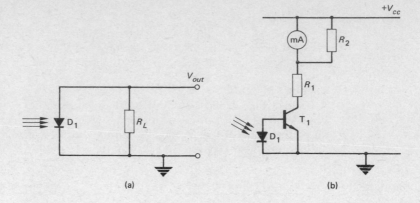

Fig. 6.7 (*a*) The photo-diode used to generate a photo-voltaic voltage, (*b*) improved circuit

the illumination in lux. The circuit used is shown in Fig. 6.7*a*. The load resistance R_L should be much larger than the ac resistance of the photo-diode. Very often the photo-voltaic voltage needs amplification before it can be accurately measured, and one possible circuit is shown in Fig. 6.7*b*. The photo-voltaic voltage is directly applied to the base/emitter terminals of the transistor and then amplified. The values of the resistors R_1 and R_2 are selected to give full-scale deflection on the milliammeter for a required incident illumination.

Photo-Voltaic Cells

Photo-voltaic cells, or **solar cells,** are essentially photo-diodes that have been constructed to produce the maximum possible output power. Low-current photo-diodes are mounted in transistor-type packages with a "window" on the top. Devices manufactured for use as solar cells are constructed to provide the maximum possible surface area.

Fig. 6.8 (*a*) Construction of a photo-voltaic cell, (*b*) typical photo-voltaic cell output characteristics

Fig. 6.8*a* shows the cross-section of a typical p-n junction solar cell. The surface layer of n-type silicon is very narrow, typically about 3×10^{-7} m, so that light energy incident upon the surface can penetrate as far as the p-n

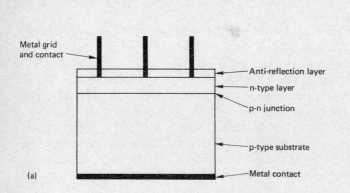

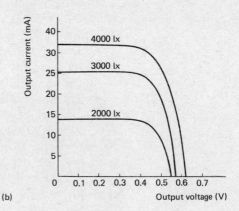

junction. A *very* thin layer is deposited on top of the n-type region whose purpose is to prevent any reflection of the incident light energy. Metal contacts, one often in the form of a grid, are applied to both sides of the cell to which input leads can be attached. Typical output characteristics for a photo-voltaic cell are shown by Fig. 6.8*b*. The output voltage can be determined by drawing a load line on the characteristics.

Photo-transistors

A **photo-transistor** is a bipolar transistor with normal collector and emitter regions but *without* a base terminal. With the base open-circuited, the collector current of a common-emitter bipolar transistor is the collector leakage current

$$I_{CEO} = I_{CBO}(1 + h_{FE})$$

I_{CBO} is the leakage current when the transistor is connected with common base and this current is due to the minority charge carriers. When the base region of the photo-transistor is illuminated, the incident light energy will generate more minority charge carriers and I_{CBO} will increase. This will give a much greater increase in I_{CEO} which, of course, means that the photo-transistor is much more sensitive than the photo-diode. On the other hand the photo-diode is much faster to operate than the photo-transistor.

Fig. 6.9 (*a*) Symbol for a photo-transistor, (*b*) photo-transistor output characteristics, (*c*) spectral response

The current/voltage characteristics of a photo-transistor are very similar to the output characteristics of a bipolar transistor, differing in that curves for different illuminations replace the curves for different base currents. Fig. 6.9*a* shows the symbol for a photo-transistor, Fig. 6.9*b* gives a set of typical *I/V* characteristics, and Fig. 6.9*c* shows a typical spectral response curve.

(a)

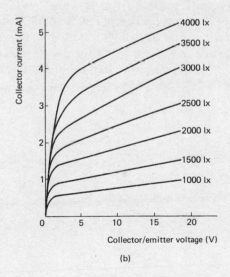

(b)

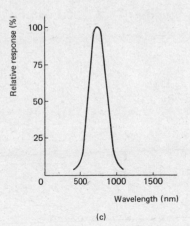

(c)

The photo-transistor can be operated in the usual common-emitter configuration (Fig. 6.10a) or as an emitter follower (Fig. 6.10b). Some photo-transistors are designed for use in conjunction with a light-emitting diode (led) as shown by Fig. 6.10c. An optically-coupled circuit like Fig. 6.10c provides a high degree of isolation between the input and output terminals, a feature which is sometimes of great importance. Some devices, known as *opto-couplers*, incorporate both the led and the photo-transistor inside the one package.

Fig. 6.10 Photo-transistors connected (*a*) in common-emitter circuit, (*b*) as an emitter follower, (*c*) in an optically-coupled circuit

For increased sensitivity and/or higher output currents the photo-transistor may be connected as a Darlington pair [EIV] with an ordinary transistor. Another alternative is the use of a photo-junction fet.

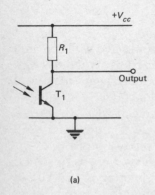 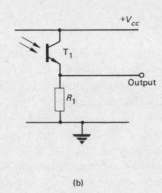 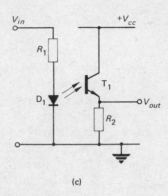

(a)   (b)   (c)

Light Transducers

A variety of other devices are available that have the ability to convert light energy into some other form of energy. Examples of such transducers are light-operated switches and opto-couplers.

The **light-operated switch** is a device that will trigger when the intensity of the incident light exceeds some particular value and will then turn ON, or OFF, a current of perhaps several amperes. Essentially, the device consists of a p-n-p/n-p-n transistor that has all its terminals externally accessible (see Fig. 6.11a). Fig. 6.11b shows the symbol for the device.

Usually, the anode gate terminal is left unconnected, the anode is con-

Fig. 6.11 (*a*) Light-operated switch, (*b*) its symbol

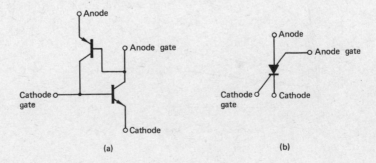

(a)   (b)

nected to $+V_{cc}$ volts via a suitable load resistance (which may be a relay coil), and the cathode is earthed. The cathode gate is connected to earth via a resistor R (or a capacitor or a R and C in parallel). When the device is illuminated, the p-n-p transistor will conduct a photo-current and this will develop a voltage across R. When this voltage reaches a value of about 0.3 V, the n-p-n transistor will turn ON and a large current will flow. The higher the resistance value chosen, the lower the incident illumination needed to trigger the device. Once the switch has been operated it can be turned OFF by applying a negative voltage pulse to the anode terminal or by switching off the supply voltage V_{cc}.

Fig. 6.12 Opto-isolator

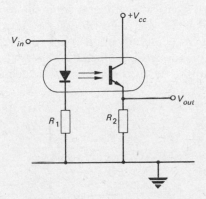

An **opto-isolator,** or **photo-coupler,** consists of a gallium arsenide light-emitting diode and a n-p-n photo-transistor fabricated inside the same case. The device has a high current transfer ratio (photo-transistor current over led current, expressed as a percentage), and, in particular, a high isolation voltage between the input and output terminals. The basic circuit is shown in Fig. 6.12. When a voltage is applied to the input terminal of the circuit, a current flows and the diode emits light energy and illuminates the photo-transistor. The transistor then conducts and the output voltage is developed across R_2. If pulses of current are passed through the led, the opto-isolator will act as a switch, the transistor conducting current for the duration of each pulse.

Although a coupled transistor has been shown in Fig. 6.12 another output device could be used, for example a silicon controlled rectifier.

Visual Displays

Visual displays are often employed in electronic equipment to indicate the numerical value of some quantity, e.g. digital watches, electronic calculators, and digital voltmeters. A variety of display devices are available but in this book the discussion will be limited to just two; namely, the light emitting diode or led, and the liquid crystal display or lcd. The circuitry needed to drive the display can be constructed using ssi gates but msi decoder/driver ics are commonly employed, sometimes in the same package as the leds.

The Light-emitting Diode

Fig. 6.13 LED symbol

The majority of **light-emitting diodes** or **leds** are either gallium arsenide (GaAs) or gallium-arsenide-phosphide (GaAsP) devices. The led radiates energy in the visible part of the spectrum when the forward bias voltage applied across the diode exceeds the voltage that turns it ON. This voltage is approximately 1.2 V for a GaAs device and 1.6 V for a GaAsP device. Four colours are readily available, red, orange, yellow and green, but blue leds are now also possible and are expected to become commercially available in the near future. The symbol for a led is shown by Fig. 6.13.

The current in a led must not be allowed to exceed a safe figure, generally some 20–40 mA, and if necessary a resistor of suitable value must be connected in series with the diode to limit the current.

Often a led is connected between one of the outputs of a ttl device and either earth or +5 V depending upon when the led is required to glow visibly. If, for example, an led is expected to glow when the output to which it is connected is *low*, the device should be connected as shown in Fig. 6.14*a*.

Fig. 6.14 Use of leds as indicators

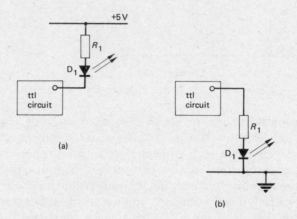

Suppose the low voltage to be 0.4 V and the sink current to be 16 mA. Then if the led voltage drop is 1.6 V the value of the series resistor will be

$$(5 - 1.6 - 0.4)/16 \times 10^{-3} = 188 \text{ ohms}$$

When the output of the device is high ($\simeq$5 V), no current flows and the led remains dark. When the led is to glow to indicate the high output condition, the circuit shown in Fig. 6.14*b* must be used. Now

$$R_1 = (5 - 1.6)/16 \times 10^{-3} = 213 \text{ ohms}$$

When a led is reverse biased it acts very much like a zener diode [EII] with a low breakdown voltage ($\simeq$4 V).

Light-emitting diodes are commonly used because they are cheap, reliable, easy to interface, and are readily available from a number of sources.

Fig. 6.15. 7-segment display: (*a*) arrangement of leds, (*b*) indication of numbers 0 to 9

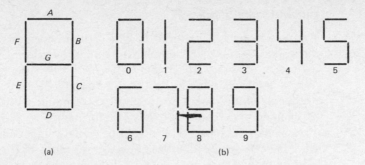

(a)   (b)

Their main disadvantage is that their luminous efficiency is low, typically 1.5 lumens/watt.

7-segment displays are generally used as numerical indicators and consist of a number of leds arranged in seven segments as shown in Fig. 6.15*a*. Any number between 0 and 9 can be indicated by lighting the appropriate segments. This is shown by Fig. 6.15*b*. A typical 7-segment display is manufactured in a 14-pin dil package with the cathode of each led being brought out to a terminal together with the common anode.

Clearly, the 7-segment display needs a 7-bit input signal and so a *decoder* is required to convert the digital signal to be displayed into the corresponding 7-segment signal. Decoder/driver circuits *can* be made using ssi devices but more usually a purpose-built ic would be used. A commonly employed device is the ttl 7447 bcd-to-7-segment decoder/driver. This ic has four input pins A, B, C and D to which the bcd input signal is applied and seven output pins, labelled as *a* through to *g*. When an output is high the segment to which it is connected lights. If, for example, the input signal is 0101, outputs *a*, *f*, *g*, *c*, and *d* go high so that the decimal number 5 is illuminated.

The ic includes a facility known as *remote blanking*. Two other pins, labelled as RB_{in} and RB_{out} are provided. If the RB_{in} pin of the most significant 7-segment display is earthed *and* inputs A, B, C and D are all low, then its RB_{out} pin will be low and so are all segment outputs. The RB_{out} pin of each display is connected to the RB_{in} pin of the next most significant 7-segment display. This connection ensures that leading 0s in a displayed decimal number are not visible, e.g. for a 4-bit display 617 would be displayed and not 0617.

Other bcd-to-7-segment decoders are available in the ttl logic family, such as the 7446/8/9 and the 74246/7/8/9, while the cmos family includes the 4511.

When a number of digits are to be displayed, as in a digital meter for example, time division multiplexing is often used to reduce the power consumption. With tdm each digit of the display is only energized for a fraction of the time the number is displayed. In a 6-digit display, for example, each digit is energized for 1/6th of the time. The disadvantage of the tdm system is the extra circuitry that is needed.

Liquid Crystal Displays

A solid crystal is a material in which the molecules are arranged in a rigid lattice structure. If the temperature of the material is increased above its melting point, the liquid that is formed will tend to retain much of the orderly molecular structure. The material is then said to be in its *liquid crystalline phase*. There are two classes of liquid crystal known, respectively, as *nematic* and *smetic* but only the former is used for display devices.

A nematic liquid crystal does not radiate light but, instead, it interferes with the passage of light whenever it is under the influence of an applied electric field. There are two ways in which the optical properties of a liquid crystal can be influenced by an electric field. These are called *dynamic scattering* and *twisted nematic*. The former was commonly employed in the past but now its application is mainly restricted to large-sized displays. The commonly-met liquid crystal displays, e.g. those in digital watches and hand calculators, are all of the twisted nematic type.

Fig. 6.16 (*a*) A liquid crystal cell, (*b*) and (*c*) operation of a liquid crystal cell

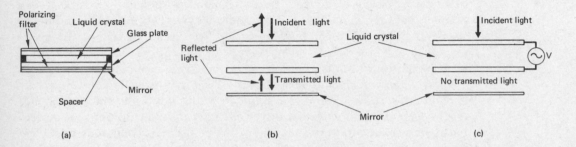

(a)   (b)   (c)

The construction of a **liquid crystal** cell is shown in Fig. 6.16*a*. A layer of a liquid crystal material is placed in between two glass plates that have transparent metal film electrodes deposited onto their interior faces. A reflective surface, or mirror, is situated on the outer side of the lower glass plate (it may be deposited on its surface). The conductive material is generally either tin oxide or a tin oxide/indium oxide mixture and it will transmit light with about 90% efficiency. The incident light upon the upper glass plate is polarized in such a way that, if there is *zero* electric field between the plates, the light is able to pass right through and arrive at the reflective surface. Here it is reflected back and the reflected light travels through the cell and emerges from the upper plate (Fig. 6.16*b*). If a voltage is applied across the plates (Fig. 6.16*c*) the polarization of the light entering the cell is altered and it is then no longer able to propagate as far as the reflective surface. Thus no light returns from the upper surface of the cell and the display appears to be *dark*.

Liquid crystal displays, unlike leds, are not available as single units and are generally manufactured in the form of a 7-segment display. The metal oxide film electrode on the surface of the upper glass plate is formed into the shape of the required 7 segments, each of which is taken to a separate contact, and the lower glass plate has a common electrode or *backplate* deposited on it. The idea is shown by Fig. 6.17. With this arrangement a

Fig. 6.17 LCD 7-segment display

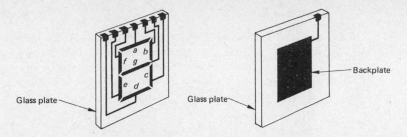

voltage can be applied between the backplate and any one, or more, of the seven segments to make that, or those, particular segment(s) appear to be dark and thereby display the required number.

Nematic liquid crystal displays possess a number of advantages which have led to their widespread use in battery-operated equipment. Firstly, their power consumption is very small, about $1 \mu W$ per segment (much less than the led); secondly, their visibility is not affected by bright incident light (such as sunlight); and, thirdly, they are compatible with low-power pmos/cmos circuitry.

Fig. 6.18 Operation of an lcd 7-segment display

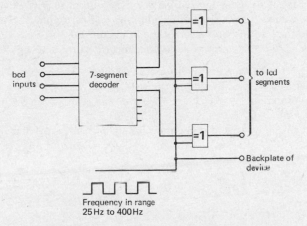

The voltages applied to a lcd must be alternating with zero d.c. component to prevent any electrolytic plating action taking place that would damage the device. The basic arrangement used is shown in Fig. 6.18; suitable lcd decoder/drivers are provided in the cmos family, e.g. the 4055 and the 4056.

Integrated circuits are also available that incorporate photo-electric devices. Examples are the Texas TIL 306/7. These devices include a bcd counter, a 4-bit latch, and a 7-segment decoder and led driver, all in a 16-pin package. The circuits can be used in applications where the clock pulse count is to be displayed, the choice of circuit being determined by a requirement for a decimal point to appear either' side of the displayed number. Several devices can be connected together so that larger numbers can be displayed.

Similar ics are also available for the display of bcd data, or 4-bit binary data, or hexadecimal data.

Exercises 6

6.1   A photo-diode is operated as a photo-voltaic device in the light-measuring circuit of Fig. 6.7b. The I/V characteristics of the diode are given by the data of Table 6.1.

Draw the diode characteristics. If the transistor has a mutual conductance of 33 mS, and fsd for the meter is 100 mV at 1 mA, determine values for R_1 and R_2 to give fsd for an illumination of (i) 1000 lx, (ii) 300 lx.

Table 6.1

Forward voltage (mV)		0	100	200	300
Reverse diode current (μA)					
Illumination	500 lx	20	19	16	0
	1000 lx	40	38	35	20
	2000 lx	70	68	66	46
	3000 lx	100	100	96	80

6.2   An led is to be connected between the output of a ttl device and +5 V. Determine the value of the series current-limiting resistor required. The led ON voltage is 1.6 V and the low output voltage of the ttl circuit is 0.2 V. Find also the power dissipated in the series resistance. Assume the safe led current to be 20 mA.

6.3   Fig. 6.19a shows a light-operated relay circuit and Fig. 6.19b gives the I/V characteristics of the photo-transistor. The relay coil has a resistance of 7500 Ω and the minimum current that will operate the relay is 1 mA. Transistor T_1 has $V_{BE(sat)} = 0.6$ V and $h_{FE} = 100$.

Determine (i) the minimum illumination intensity that will operate the relay, (ii) the voltage drop across the photo-diode when the relay operates.

Fig. 6.19

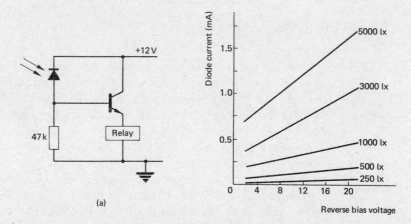

(a)

6.4   A photo-conductive cell has a resistance/illumination characteristic given by Table 6.2. It is connected in series with a resistance of 1000 Ω across a d.c. voltage supply of 10 V. Determine the incident illumination when the voltage across the cell is (i) 5 V, (ii) 2 V.

Table 6.2

Illumination (lx)	10	100	400
Resistance (Ω)	5000	600	200

6.5 A photo-coupler has the characteristics given in Figs. 6.20*a* and *b* and it is used in the circuit shown in Fig. 6.12. Determine the voltage across R_2 when $R_2 = 1000 \, \Omega$ if the input voltage is 1.4 V and $R_1 = 47 \, \Omega$ and $V_{cc} = 25$ V.

Fig. 6.20

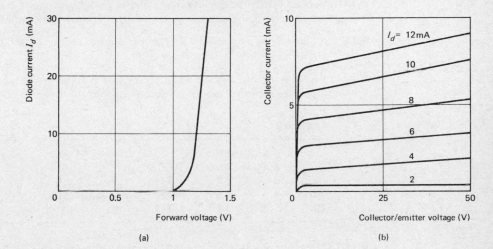

(a)　　　　　(b)

6.6 Write down the truth table for a bcd-to-7-segment decoder, assuming a luminous segment is illuminated by the logical 1 state. Use it to obtain Boolean expressions for the decimal numbers 5 and 9.

Short Exercises

6.7 Explain the operation of the circuit given in Fig. 6.21 and suggest a suitable application for it.

Fig. 6.21

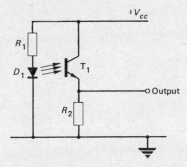

6.8 Draw a diagram to show how four 7447 bcd/7-segment decoder/drivers could be connected to give remote blanking.

6.9 Answer the following questions relating to an lcd display:
   a) Can the display be read in the dark?
   b) Will the display work in very low ambient temperatures?
   c) Is the display affected if it is situated in very bright sunlight?
   Give reasons for your answers.

6.10 Describe the principle of operation of a photo-voltaic cell.

6.11 What is meant by multiplexing when applied to a 6-digit led display? State one disadvantage and one advantage of using multiplexing.

6.12 Explain briefly the operation of a photo-diode. What are the effects of varying (i) the wavelength of the incident radiation, (ii) the ambient temperature?

6.13 Explain how a photo-diode can be used as either a photo-conductive cell or a photo-voltaic cell.

6.14 An led is used to indicate when a 12 V power supply voltage is present. Calculate the necessary series resistance for the diode current to be 25 mA.

Table 6.3

V_{CE} (V)	Illumination		
	3000 lx	2000 lx	1000 lx
5	3.8	2.2	0.9
10	4.3	2.5	0.98
20	5.2	3.1	1.0

(Diode current limit)

6.15 A photo-transistor has the I/V characteristic given by Table 6.3. The transistor is connected in series with a relay of resistance 3000 Ω and a supply voltage of 15 V. The relay will operate when the current flowing in its coil is 2 mA. Estimate the smallest illumination that will operate the relay.

7 Memories

Introduction
Many digital systems include some form of *memory* or *storage* where data can be held on either a permanent or a temporary basis. A number of magnetic devices, such as discs and tapes, are available for use as permanent stores and the ferrite core store has been popular as a temporary store. Magnetic stores possess the advantage of being **non-volatile;** this term means that they are able to retain stored data when the power supplies have been turned off. Semiconductor memories have now become the predominant memory technology, particularly for short-term storage, and their capacity is being increased yearly as improved techniques are developed. The reasons for the increasing use of integrated circuit memories are that they are smaller, cheaper, and faster to operate than the magnetic alternatives.

Only **semiconductor memories** are to be discussed in this chapter but a discussion of ferrite core memories can be found in a preceding volume [DT&S].

A memory consists of a matrix of memory **cells** and a number of digital circuits that provide such functions as *address selection* and *control*. Each cell has a particular location in the matrix that is identified by a unique **address**.

Three kinds of semiconductor memory are used, known as the random access memory or ram, the read only memory or rom, and the serial access memory. The serial access memory is generally provided by a shift register and these circuits have been considered in Chapter 5. Charge-coupled devices are also serial memories.

The *random access memory* or *ram* can have data read out of it, or written into it, with the same access time for all addresses. The **access time** is the time that elapses between the start of a memory request and the data becoming available. Both *dynamic* and *static* rams are possible, with the former being both cheaper and simpler to fabricate but possessing the disadvantage that the stored information must be periodically *refreshed*. On the other hand, the dynamic memory is faster and allows a greater packing density to be achieved.

The other kind of memory is the *read only memory* or *rom* and it can only be used to store permanent data. Some types of rom are programmed by the semiconductor manufacturer and can only be used to store the data designed into the device. Other types, known as *programmable read only memories* or *proms,* can be set up for a particular purpose after the equipment manufacturer has taken delivery of the memory.

The **capacity** of a memory is measured in terms of *words* and *bits,* with the total number of bits being quoted as so many K. The letter K does *not*

149

denote 1000 as is the custom with resistors but it represents 2^{10} or 1024. Thus a 1K store is able to hold 1024 bits of information. This capacity could be arranged in a number of different ways: for example, 256 4-bit words or 128 8-bit words or 1024 1-bit words. In all cases the number m of words is equal to 2^n and this means that any word can be addressed by n suitably decoded bits. Thus, 256 words can be addressed by 8 input lines.

Many memories store only 1-bit words because there is then only a need for *one* input/output data terminal. This is particularly true for the larger memories since the number of input and output terminals that can be provided is limited by the number of ic package pins that are available. It also means that only a one-dimensional address is needed to identify each location.

The standard memory sizes are 1K, 4K, 16K, 32K, and 64K. CMOS rams are currently available with densities ranging from 1K to 16K. Memories can be connected in *arrays* of ×1, ×4, and ×8 to obtain almost any required memory capacity.

Random Access Memories

A semiconductor **random access memory** is arranged in the form of a matrix of m words times n bits with the memory elements or *cells* at the intersections (see Fig. 7.1a). Although a square matrix has been drawn in the figure and is often used, e.g. $32 \times 32 = 1024$ bits, other matrices are also employed, e.g. $64 \times 16 = 1024$ bits. Each memory cell has an individual address in the matrix and can be selected by the simultaneous application of voltages to one row and one column address line.

To reduce the number of address input lines necessary, addresses are inputted in binary and decoded as shown by Fig. 7.1b. The ic package must have the following pins:

Input and output data bits.
Address bits.
A read/write bit.
A chip select (or chip enable) bit.

A 64×4-bit memory for example, would require

a) 6 address pins since $2^6 = 64$.
b) 4 input data pins.
c) 4 output data pins.
d) 1 read/write pin.
e) 1 chip enable pin.

This is a total of 16 pins to which must be added pins for $+V_{cc}$ and earth connections. Thus, an 18-pin dil package is required.

If the same memory capacity is arranged as 1024 1-bit words the package pins required are

a) 10 address pins since $2^{10} = 1024$.

b) 1 data input pin.

c) 1 data output pin.

d) 1 read/write pin.

e) 1 chip select pin.

Fig. 7.1 (*a*) Basic arrangement of a ram matrix, (*b*) block diagram of a ram

These, plus pins for $+V_{cc}$ and earth, can be accommodated in a 16-pin dil package.

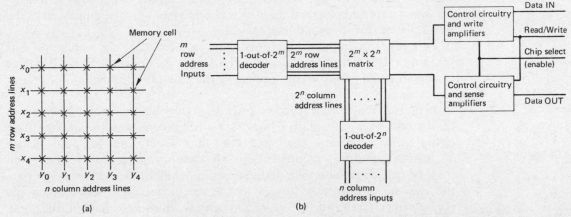

(a)

(b)

The chip select or the chip enable input is used to select a particular memory from a number that are connected in parallel in order to produce a larger capacity memory. The data output lines are either open-collector (p. 27) or three-state. A **three-state output** can be placed into any one of three distinct states; it can be in either of the logical 0 or 1 states or it can be switched to have a very high output impedance. The third condition, in which the output is generally assumed to be open-circuited, is applied when the chip is disabled (chip select input at 1) and makes it possible for memory chips to be paralleled to form larger memories. (A three-state output is often known as *tri-state*.)

When the logical state of a location in a memory is to be **read,** the read/write line is set to the READ level (usually logical 1) and the address of the required location is fed in. The data held at that location is presented at the data output terminal(s) and there it remains until such time as any one, or more, of the address, read/write or enable inputs change state. The stored data is *not* affected by the read-out process; in other words the read-out is *non-destructive*.

When data is to be **written into** the memory, the read/write line is set to the WRITE level (usually logic 0). Then data present at the input data line(s) will be entered into the memory at the addressed location. Any data that was already at that location is *lost*. This does mean, of course, that the input data must not change whilst the chip is in the write state.

The timing of a memory operation or *cycle* is critical. The time for a READ cycle to be executed is limited by the **access time** of the memory. This is the time taken for data to be read out of the memory after the chip select, read/write, and address bits are steady. The time for a WRITE cycle to be executed is limited by the *write* or cycle time. This is the time for which the address, chip select, read/write and input data bits must be held constant in order to correctly write the new data into the addressed location. Very often the access and the cycle times are equal, e.g. for the Mostek $1K \times 8$ static ram the access and cycle times are both equal to 200 ns.

There are two main classes of ram manufactured, namely *static rams* and *dynamic rams*. Static rams can be manufactured using either bipolar transistor or mosfet technology but only mosfet techniques can be employed in the manufacture of a dynamic ram.

Very often the number of words to be stored and/or the number of bits per word is larger than the capacity of a single memory chip. A larger capacity memory can be constructed by the suitable combination of two or more chips.

A 1K ram may be arranged to provide a $1K \times 1$ bit memory. This means that it will have ten address lines ($2^{10} = 1024$), one data input, one data output, one read/write line, and one chip select line. Suppose that a $1K \times 4$-bit memory is required. Four 1K memory chips are necessary and must be connected with their address pins in parallel. The read/write pins and the chip select pins must also be connected in parallel The necessary pin connections are shown in Fig. 7.2. When the chips are enabled, the same location is selected in each memory and the data can be read out of pins D_4, D_5, D_6 and D_7 or, if the WRITE mode has been selected, written into pins D_0, D_1, D_2 and D_3.

Fig. 7.2 Four $1K \times 1$ rams connected to provide a $1K \times 4$ ram

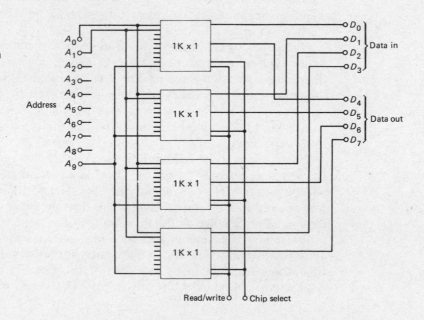

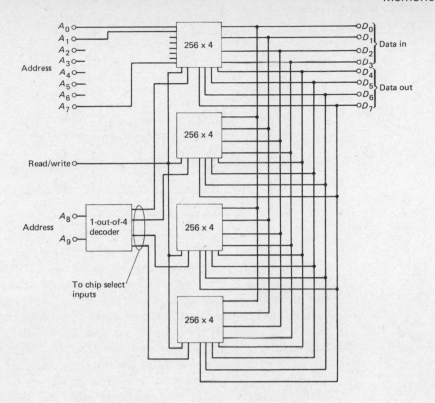

Fig. 7.3 Four 256×4 rams connected to provide a $1K \times 4$ ram

The number of words that a memory can store can be increased by combining two or more chips in the manner shown in Fig. 7.3. Here four 256×4-bit (again 1K) chips are connected with their address, $\overline{\text{read}/\text{write}}$, and input/output data lines in parallel. The chip select inputs of the four chips are connected to the four outputs of a 1-out-of-4 decoder.

The address lines A_0 through to A_7 select the required memory cell in each of the four memory matrices but *only one* of the chips is enabled by the 2-bit address A_8/A_9 applied to the decoder. Thus, one only address at a time in the combined memory can be read out of or written into.

Static Random Access Memories

A **static ram** consists of a matrix of flip-flops each of which is able to store one bit of information and acts as a memory cell. Static rams can be fabricated using either bipolar transistor or mosfet technology. Bipolar transistor rams are mainly in the ttl family although ecl and I^2L devices are also available, while mosfet devices are either nmos or cmos.

Fig. 7.4 TTL memory cell

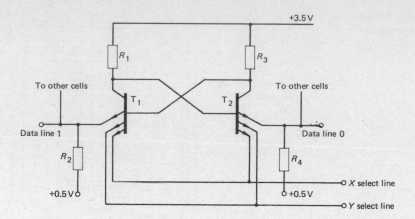

The circuit of a **ttl memory cell** is shown in Fig. 7.4. Transistors T_1 and T_2 are connected as a flip-flop. The collector supply voltage is shown as +3.5 V but may be any value up to +5 V. This means that logic 0 is represented by voltages below about +0.3 V and logic 1 by voltages above about +3 V. When the cell is in the state T_1 ON and T_2 OFF, the circuit stores a 0 bit. Conversely, a 1 bit is stored by the state T_1 OFF and T_2 ON. An n-p-n transistor with multiple emitters will turn OFF when *all* its emitters are held at a voltage that is more positive than the base potential. If any one or more of the emitters are at approximately 0 V, and the base is positive, the transistor will conduct no matter how positive the other emitters should be.

Suppose that both the X and Y lines are at logic 0 and that the logic state of the cell is T_1 ON and T_2 OFF. The current conducted by T_1 will flow out of the two emitters connected to the X and the Y lines but no current flows into the data 1 line because this emitter is held at a voltage greater than the 0 logic level. If *either* the X *or* the Y line is addressed so that its voltage rises to the logic 1 level, there will be no change in the operation of the circuit since one emitter of each transistor will still be low.

When the cell is addressed by *both* the X and Y lines having a logic 1 voltage level applied, the associated emitters become more positive than the "data" emitter. The current conducted by the ON transistor now flows out of the cell into the data line. This current then flows into circuitry not shown in the figure and it causes the data output terminal to assume the same logic level as the cell.

When data is to be written into an addressed cell, either the data 1 line or the data 0 line is taken to the logic 1 voltage level to turn OFF the associated transistor (all three emitters are then high). A cell that has not been addressed will not pass current to the appropriate data line and does not respond to the write operation because each transistor will have at least one emitter at the logic 1 voltage level.

The circuit of the **Schottky transistor memory** cell is shown in Fig. 7.5. When the cell is not addressed, line X is at the logic 1 voltage level or at about +2.5 V. This means that the two diodes D_1 and D_2 are reverse-biased

Fig. 7.5 Schottky transistor memory cell

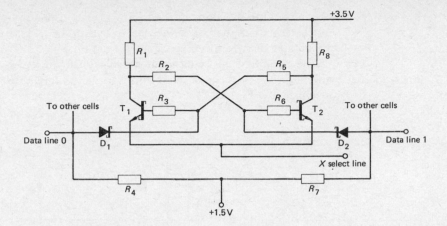

and so non-conducting. Suppose that the flip-flop is in the state T_1 ON and T_2 OFF and that the cell is addressed by the voltage on line X being reduced to the logic 0 level of about +0.3 V. The base potential of T_1 will then be equal to $(+0.3 + V_{BE})$ volts $\simeq 1.0$ V and the diode D_1 will become forward biased by approximately 0.5 V. The voltage on the data 0 line will fall from +1.5 V to $1.0 + 0.4 = 1.4$ V since the voltage drop across a Schottky diode is about 0.4 V. The collector potential of T_1 is now equal to 0.3 V $+ V_{CE(sat)} \simeq 0.5$ V and so the voltage on the data 1 line is much greater than the voltage change on the data 0 line. The *difference* between the two voltage changes is used as an indication of the logical state of the cell. Only this cell is read since all the other cells will have their X line high so that their diodes D_1 and D_2 are OFF.

When data is to be written into a cell, the voltage of the X line is reduced to 0.3 V to turn either D_1 or D_2 ON. Suppose that as before the cell is in the state T_1 ON and T_2 OFF and that the other state is required. The voltage of the data 1 line is then increased to about +2.8 V to turn transistor T_2 ON. The normal flip-flop action then ensures that T_1 rapidly turns OFF. If the required state had been T_1 ON and T_2 OFF, the +2.8 V voltage would have been applied to the data 0 line and this would, of course, have had no effect on the logical state of the cell. Non-addressed cells are not written into since their X line voltage remains at +2.5 V and so the collector potential of their ON transistor is $2.5 + 0.2 = +2.7$ V. Thus a data line voltage of $+2.7 + 0.4 = +3.1$ V is needed to initiate switching.

The Schottky transistor memory cell possesses advantages over the multiple-emitter transistor cell in that it is faster to operate and it dissipates less power. On the other hand it can be addressed in the X plane only.

Some of the memories in the ttl family are: 7489/74170/74LS170 (4×4), and the 74LS670 (4×4). The first two have a totem pole output, while the last ram has a three-state output. TTL memories are of relatively small capacity, and are mainly employed as fast, small temporary stores. When larger storage capacities are needed a mosfet circuit must be used. This is because mosfet circuitry allows much greater packing densities to be achieved so that large capacity memories can be fabricated within a single chip. An added advantage is that the power consumption of a mosfet cell is much less than that of a bipolar transistor cell. The price paid for these advantages is a reduced speed of operation.

The circuit of an **nmos memory cell** is shown in Fig. 7.6. Transistors T_3 and T_5 are connected as a flip-flop with active drain load resistances provided by T_2 and T_4.

Fig. 7.6 NMOS memory cell

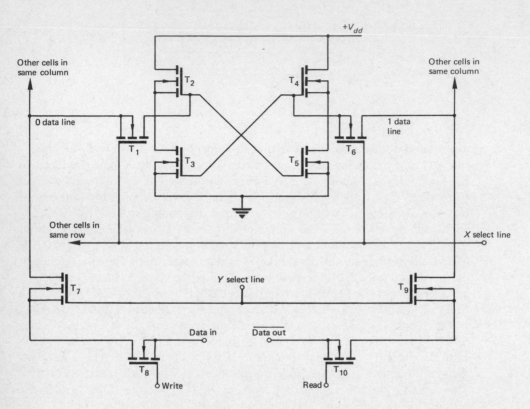

Transistors T_1 and T_6 form gates which open to connect the cell to the data lines when the cell is addressed. Six transistors form one memory cell. Transistors T_7, T_8, T_9 and T_{10} form a part of the control circuitry and are common to all the cells in the same column of the memory matrix. Note that all the transistors are n-channel enhancement-mode mosfets that are able to act as bi-directional switches.

The memory cell is addressed by making both the X and the Y select lines go high (approximately $+V_{dd}$ volts). The X select voltage turns transistors T_1 and T_6 ON and connects the cell to the 0 and 1 data lines. The Y select voltage turns T_7 and T_9 ON to connect the write and read circuits to the selected column and hence to the addressed cell.

To read the state of a cell, the READ terminal is taken high, turning the transistor T_{10} ON. The drain potential of T_5 then appears, with little voltage drop, across T_6, T_9 and T_{10}, at the data-out terminal. For new data to be written into the cell, the WRITE terminal is set to the logical 1 voltage level.

This turns ON transistor T_9 and applies the DATA IN voltage to the gate terminal of T_5. If the input data is 1, transistor T_5 turns ON and T_3 turns OFF so that the drain potential of T_5 is approximately 0 V. Conversely, if the input data is 0, then T_5 turns OFF and T_3 turns ON and then the drain potential of T_5 is high.

The operation of the circuit is such that the state T_3 OFF and T_5 ON indicates that the stored bit is at logical 1 and vice versa. Hence the read-out from the circuit is inverted, as indicated by the labelling of the terminal in the figure as $\overline{\text{Data out}}$.

Fig. 7.7 CMOS memory cell

Bipolar transistor cells are also fabricated using ecl and I²L logic.

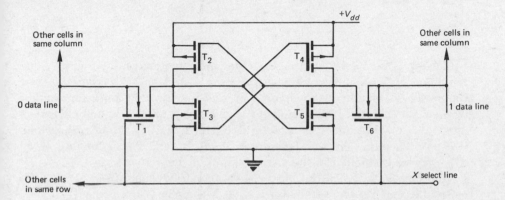

Fig. 7.7 shows the circuit of a **cmos memory cell.** The pairs of transistors forming the opposite sides of the flip-flop consist of one p-channel and one n-channel mosfet. The use of complementary mosfets in this way increases the speed of operation and reduces the power dissipation. As in the nmos circuit, mosfets T_1 and T_6 connect the cell to the data lines when the X select line is high. The bit stored is 1 when the circuit is in the state T_3 OFF and T_4 ON and 0 when it is T_3 ON and T_4 OFF.

MOS memories are of larger capacity than the bipolar transistor alternatives. For example, the 2102 is organized as a 1024×1-bit ram with an access time of 300 ns. The pin connections of this three-state device are shown in Fig. 7.8. Other memory sizes are $16\,384 \times 1$-bit, 4096×1-bit words, and 256×4-bit words. CMOS memories are considerably slower to access than other memories.

CMOS rams can be made virtually non-volatile by the provision of a self-contained battery supply that can be switched into circuit should the power supply fail. Since cmos devices dissipate very little power only a small capacity battery is required, say 0.4 Ah.

Dynamic Random Access Memories

A reduction in the number of transistors needed per cell can be achieved if a **dynamic memory** is employed. In this type of memory data is stored in the stray capacitances which exist between the gate and the source terminals of a mosfet.

Fig. 7.8 Pin connections of the 2102 1024×1 ram

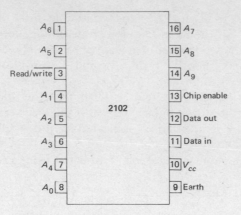

One type of dynamic cell that uses **four mosfets** is shown by Fig. 7.9. The reduction in the number of transistors from 6 to 4 does, of course, reduce the power dissipation of the circuit, but it also means that a greater number of memory cells can be packed into a given chip area. The cell comprises mosfets T_3 and T_7 connected as a flip-flop with T_2 and T_6 acting as their load resistors. The other mosfets, labelled as T_1, T_4, T_5 and T_8 are common to all the other cells situated in the same column of the matrix.

The logical state of a cell is stored by the stray capacitances C_1 and C_2 across the gate/source terminals of T_3 and T_7. When transistors T_2, T_4, T_6 and T_8 are turned ON by voltages applied to the X and Y select lines, the capacitances are effectively connected to the two data lines.

When the cell is storing the logical 1 state, T_3 is OFF and T_4 is ON. At this time stray capacitance C_1 has zero voltage across it while there is a positive voltage across the stray capacitance C_2.

To read the logical state of the cell it must first be addressed by making both the X and Y select lines go to the logical 1 voltage level. The voltage at the drain of T_7 is then fed to the data out terminal via transistors T_6 and T_8. Note that once again the logic 1 state is read as a low voltage.

New data is written into the cell by applying the appropriate voltage to the source terminal of T_4. This voltage is applied, via T_2 and T_4, to capacitance C_2 and hence between the gate and source terminals of T_7. If this voltage is high T_7 will turn ON and T_3 will turn OFF; conversely, if this voltage is low T_7 will turn OFF and T_3 ON.

The action of writing new data into the cell replaces or *refreshes* the charge stored in either capacitance. If a WRITE operation is not carried out for some time, this charge will leak away and the stored data will be lost. For this reason it is necessary to refresh the cell at regular intervals of time. The refreshing action is accomplished by setting both the X select and the refresh terminals to the logical 1 voltage level. This action turns ON T_1, T_2, T_5, and T_6. Voltage V_{dd} is then applied to both capacitance C_1 and C_2 but one of these capacitances is in parallel with an OFF mosfet and charges up quickly. The other capacitance is in parallel with an ON mosfet and so most of the charging current passes through the transistor and does not

Fig. 7.9 Dynamic memory cell

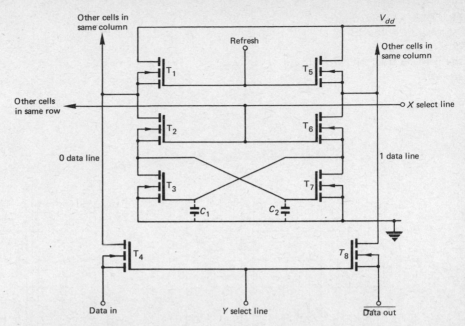

charge the capacitance. Thus only the capacitance that is already holding some charge is refreshed. The situation is illustrated in Fig. 7.10 for the logical state T_3 ON, T_7 OFF, i.e. logic 0 stored.

Other versions of the mosfet dynamic memory cell have been developed, some of which use fewer transistors, three or even one. The basic circuit of a **one-mosfet dynamic cell** is shown in Fig. 7.11. The bit to be stored is held by the capacitance C_1; zero charge corresponds to the logic 0 state and full charge to the logic 1 state. To read the state of the cell, it must first be addressed by applying a voltage to the X select line. The charge, or lack of charge, on C_1 is then fed to the Y access line and is there sensed by an amplifier and passed on to the data output terminal of the chip. The read-out process is destructive and must always be followed by the WRITE operation. Writing a bit into the cell is accomplished by applying the appropriate voltage to the Y select line. At the same time all the other cells in the same row are also refreshed.

Fig. 7.10 Refreshing a dynamic memory cell

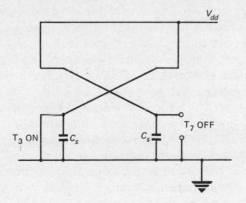

Fig. 7.11 Single mosfet dynamic memory cell

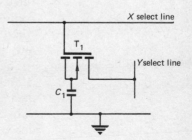

Fig. 7.12 A 9-bit ram (*a*) organized as 3×3-bit words, (*b*) organized as 9×1-bit words

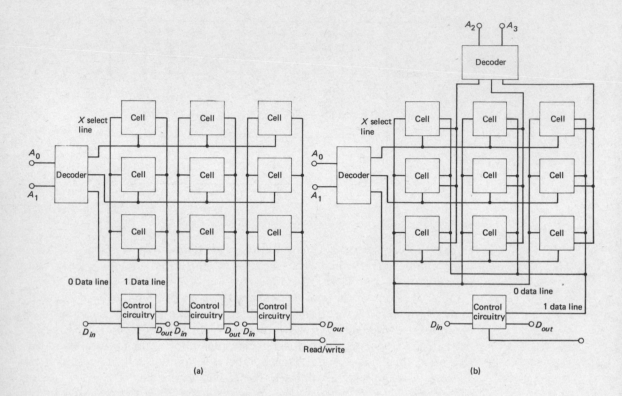

(a) (b)

Organization of a RAM

A ram consists of a matrix of memory cells that can be organized in a number of different ways. As a simple example consider a 9-bit memory. This could be organized as three 3-bit words as shown by Fig. 7.12*a*, or as nine 1-bit words as shown by Fig. 7.12*b*. For the 3×3 memory the decoded address takes one of the X select lines high to select a particular word. All three cells in the same column have common data 0 and data 1 lines and the control circuitry is such that the input data D_{in} is written into the memory when the read/write line goes high. Conversely, when the read/write line is low, the addressed word is read out of the memory.

The 9×1 memory is organized differently (Fig. 7.12*b*). Each cell is addressed in two planes and so each cell can be individually located. The data 0 and data 1 lines are fed to each cell in parallel as shown and, since only one bit at a time can be read out of, or written into, the memory, only one control circuit block is shown.

**Read Only
Memories**

Read only memories or **roms** have data permanently written into them either by the manufacturer or by the user, and can *only* be used to read out data. Data is *not* written into the memory during a circuit operation. Read only memories are used for a variety of purposes including storage of mathematical tables (such as logarithms and trigonometric functions) and code conversion and programme storage. A rom is non-volatile.

A rom is organized in a similar way to a ram in that data is stored at different locations in the memory matrix and each location has a unique address. When a particular location is addressed, the data stored at that address is read out of the memory. The read-out is non-destructive. Address decoders are used to reduce the number of input address bits required and so allow the rom to be fabricated within a standard dil package.

Fig. 7.13*a* shows the arrangement of a diode rom. Although diodes are shown, bipolar or field effect transistors could be used instead, connected in the manner shown by Figs. 7.13*b* and *c*. In any case the rom is an integrated circuit device fabricated within an ic package. The logical output of the rom is determined by the connections made by the diodes between an address decoder line, 0 through to 7, and an output line, A through to F. Normally each output decoder line is high and goes low only when it is selected by an input address. Each diode that is connected to a low decoder output line is turned ON and takes the associated output line low also.

Fig. 7.13 (*a*) Diode rom, (*b*) use of bipolar transistor, (*c*) use of mosfet

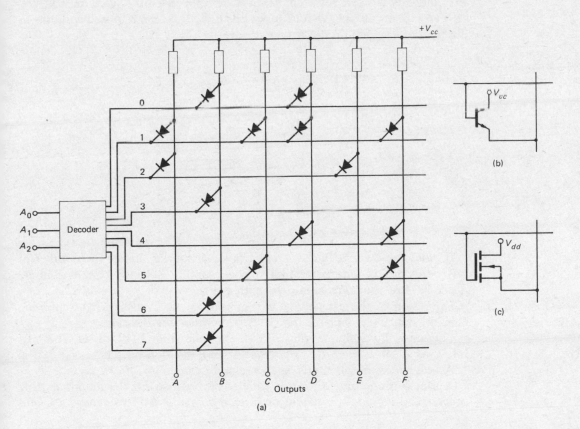

(a)

If, for example, the input address is 010, decoder output line 2 will go low. This row line has two diodes connected to it; these two diodes conduct and take the output lines A and E low also. Thus the output of the rom is 011101. If the input address is 110, decoder line 5 goes low and the rom output is 110110. The output signal will remain available for as long as the address is held.

When either kind of transistor is used as the conducting element, a decoder output going high will turn ON the associated transistors and give the logical 0 state at each of the connected outputs. Those outputs that are not linked by a transistor to the selected row will remain at logical 1.

When data is stored in a rom the number of rom outputs should correspond to the number of bits per word. If the arrangement shown in Fig. 7.13a is used as it stands this would not be the case, unless, of course, the number of columns was only four, or perhaps eight. A typical size for a rom is 4 kilobits and, if this is organized with a square matrix, there will be 64 rows and 64 columns. Clearly, some sort of output circuitry is needed since 64-bit words are not employed.

Suppose that the 4 kbit memory is organized as 1028 4-bit words. The row address decoder must have six input bits since $2^6 = 64$. The four data output bits, D_0, D_1, D_2 and D_3 must be obtained from 64 columns. Thus, the columns will be divided into 16 groups of four bits, and this means that there is a need for four 16-to-1 selector circuits. Since $16 = 2^4$ the selector circuits must have four input bits to address the required columns in the matrix. The required circuit is shown by Fig. 7.14.

Fig. 7.14 Organization of a rom

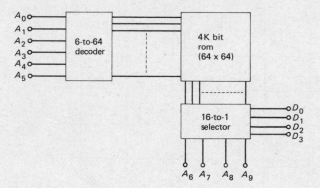

It should be evident that if the matrix is not square, and/or the number of bits per word is different, other decoder and/or selector circuits will be needed. Most roms are organized with either 4-bit or 8-bit words.

Expansion of the capacity of a rom can be accomplished by combining chips in similar manner to that described earlier for the ram.

Commercially available roms include the following: 2048×8-bit, $16K \times 8$-bit, 256×4-bit, 256×8-bit, 1024×4-bit and 1024×8-bit.

Access times are generally in the region of 60 ns to 250 ns.

A rom is programmed for a particular purpose during the manufacturing process and its function cannot be altered by the user. This can be a useful feature for an equipment manufacturer who may have a need for a large number of identical devices. However it is often much less convenient for the user of only small quantities of roms. In order to provide some flexibility in the possible applications for roms, *programmable* devices have been introduced.

Programmable ROMs

A **programmable rom** or **prom** is manufactured as a generalized integrated circuit with *all* the matrix intersections linked by *fusable* diodes or transistors (Fig. 7.15). Because the circuit *is* generalized it can be mass-produced, which, of course, brings the cost down. This means that *all* outputs will go to logical 0 whenever *any* of the decoder outputs is addressed. The purchaser of a prom can programme the device to perform for a specific application.

Programming is carried out by addressing a particular intersection by selection of its row and its column and then taking the base, or the gate, of the transistor to a high positive voltage and the column line to near earth potential. The current taken by the transistor is then large enough to blow the "fuse" and leave the emitter, or the source, of the transistor open-circuited. Each link thus destroyed will result in the associated output line remaining at logical 1 when that location is addressed.

The programmer must be very careful when carrying out this work since any links that may be incorrectly fused cannot be restored.

PROMs are widely used in the control of electrical equipment such as washing machines and electric ovens. Bipolar proms are faster and cheaper than mosfet proms but their power dissipation is higher. Also, mosfet proms can have much larger matrices.

Erasable and Electrically-alterable PROMs

A further development of the prom has led to the introduction of devices that can be re-programmed if necessary. These devices utilize a different mosfet cell structure in which programming is accomplished by the storage of a charge in the cell instead of by the blowing of a fuse. With an **erasable prom** or **eprom,** erasure of an existing programme is achieved by exposing the chip to ultra-violet radiation directed through a window in the chip package. The radiation increases the conductivity of the cell and in so doing allows the stored charge to leak away. The process cannot be applied to an individual cell but only to the whole matrix, and so the erasure procedure results in *all* the stored data being lost. After erasure, re-programming of a cell is carried out by applying voltages to the gate and drain of the mosfet and earthing the source and the substrate. This causes the cell to store the charge that indicates the logical 1 state.

Another kind of re-programmable rom is known as the **electrically-alterable prom** or **eaprom**. As with the eprom, programming a cell to store a 1 bit is done by charging the cell, but now cell erasure is accomplished by the application of a reverse-polarity voltage to the mosfet that removes the stored charge. With this kind of device individual cells can be re-programmed without affecting the other cells in the matrix.

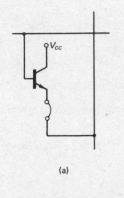

(a)

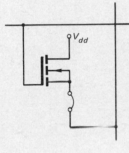

(b)

Fig. 7.15 Elements in (*a*) a ttl prom, (*b*) a mosfet prom

Exercises 7

7.1 Sketch and describe the circuit of a cell used in a bipolar transistor store. Draw a circuit to show how the cells are connected to each other to form a matrix.

With the aid of a block diagram explain how a particular location in the store is addressed.

7.2 Distinguish between coincident selection and linear selection of a memory cell and give some advantages and disadvantages of each method.

Draw the circuit of a coincident selection ram cell and explain its operation.

7.3 Design a rom to read out the squares of the numbers 1 through to 7.

7.4 Explain, with the aid of a suitable diagram, the operation of any kind of rom. Explain the difference between a rom and a prom. What are the advantages of the latter?

7.5 Draw the circuit diagram of a dynamic ram cell and briefly explain its operation. Why is periodic refreshing necessary?

7.7 Show how a number of 256×1 memories can be combined to produce a 1024×1 memory. Design the necessary decoder.

Fig. 7.16

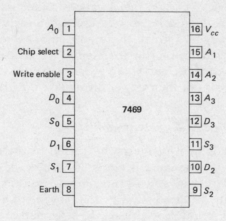

A = address, D = data in, S = sense (data) out

7.7 Fig. 7.16 shows the pin connections of the 74LS89 64-bit ram that is arranged internally as a 16×4-bit word. Draw a diagram to show how two such circuits can be combined to make a 16×8-bit memory. State, without drawing a circuit, how a 32×4-bit memory could be achieved.

Short Exercises

7.8 With the aid of a block diagram explain how a ram can be used as a code converter.

7.9 Explain what is meant by each of the following types of memory: (i) serial access, (ii) random access. Give two examples of each type of memory and state two merits of each.

7.10 Draw a diode matrix that could produce the square of the decimal numbers 1, 2, 3, and 4.

7.11 Explain the difference between dynamic and static memories.

7.12 Explain clearly the differences between a ram and a rom. State when each would be used.

7.13 Explain what is meant by the terms access time and cycle time when applied to a semiconductor memory.

7.14 List the relative advantages and disadvantages of (i) bipolar and mosfet static rams, (ii) static and dynamic rams. Why are bipolar transistor dynamic rams not available?

7.15 A 256-bit memory is arranged as 32 words of 4 bits each. How many input/output pins are needed on the dil package? Given that standard packages have an even number of pins, suggest a more efficient memory size from a pin utilization point of view.

7.16 The Schottky transistor memory cell cannot be addressed in two planes. What limitation does this place on a memory made up of such cells?

7.17 How many 7489 16×4-bit memory chips must be combined to produce (i) a 64×4-bit memory, (ii) a 64×8-bit memory?

7.18 The description of a commercially available ram is: it is a 64-bit ram organized as a 16 word $\times 4$-bit array; address inputs are buffered and fully decoded on chip; the outputs are three-state and are high impedance when the $\overline{\text{CS}}$ input is high.
  Explain briefly the meaning of each sentence.

Digital Electronics IV: Learning Objectives (TEC)

(C) Logic Gates

(D) Logic Networks

(E) Memory Circuits

(F) LSI and VLSI Memories

(G) Counters and Registers

(H)   LSI Logic Circuits

Index